www.EffortlessMath.com

... So Much More Online!

- ✓ FREE Math lessons
- ✓ More Math learning books!
- ✓ Mathematics Worksheets
- ✓ Online Math Tutors

Need a PDF version of this book?

Send email to: Info@EffortlessMath.com

Algebra 2

Workbook

A Comprehensive Review and Step-by-Step Guide for Mastering Essential Math Skills

By

Reza Nazari

& Ava Ross

Reza Nazari & Ava Ross

All inquiries should be addressed to:

info@effortlessMath.com

www.EffortlessMath.com

ISBN–13: 172292554X

ISBN–10: 978-1722925543

Published by: Effortless Math Education

www.EffortlessMath.com

Description

Algebra 2 Workbook provides students with the confidence and math skills they need to succeed in any math course they choose and prepare them for future study of Pre–Calculus and Calculus, providing a solid foundation of Math topics with abundant exercises for each topic. It is designed to address the needs of math students who must have a working knowledge of algebra.

This comprehensive workbook with over 2,500 sample questions is all you need to fully prepare for your algebra 2 course. It will help you learn everything you need to ace the algebra 2 exam.

Inside the pages of this comprehensive workbook, students can learn algebra operations in a structured manner with a complete study program to help them understand essential math skills. It also has many exciting features, including:

- Dynamic design and easy–to–follow activities
- A fun, interactive and concrete learning process
- Targeted, skill–building practices
- Fun exercises that build confidence
- Math topics are grouped by category, so you can focus on the topics you struggle on
- All solutions for the exercises are included, so you will always find the answers

Algebra 2 Workbook is an incredibly useful tool for those who want to review all topics being taught in algebra 2 courses. It efficiently and effectively reinforces learning outcomes through engaging questions and repeated practice, helping you to quickly master Math skills.

About the Author

Reza Nazari is the author of more than 100 Math learning books including:
– **Math and Critical Thinking Challenges:** For the Middle and High School Student
– **GRE Math in 30 Days**
– **ASVAB Math Workbook 2018 - 2019**
– **Effortless Math Education Workbooks**
– **and many more Mathematics books …**

Reza is also an experienced Math instructor and a test–prep expert who has been tutoring students since 2008. Reza is the founder of Effortless Math Education, a tutoring company that has helped many students raise their standardized test scores—and attend the colleges of their dreams. Reza provides an individualized custom learning plan and the personalized attention that makes a difference in how students view math.

You can contact Reza via email at:
reza@EffortlessMath.com

Find Reza's professional profile at:
goo.gl/zoC9rJ

Contents

Chapter 1: Fundamentals and Building Blocks

Topics that you'll learn in this chapter:

- ✓ Multiplying and Dividing Integers
- ✓ Order of Operations
- ✓ Scientific Notation
- ✓ Integers and Absolute Value
- ✓ Variable Expressions
- ✓ Simplifying Polynomial Expressions
- ✓ Translate Phrases into an Algebraic Statement

Multiplying and Dividing Integers

Step-by-step guide:

Use these rules for multiplying and dividing integers:

- ✓ (negative) × (negative) = positive
- ✓ (negative) ÷ (negative) = positive
- ✓ (negative) × (positive) = negative
- ✓ (negative) ÷ (positive) = negative
- ✓ (positive) × (positive) = positive

Examples:

1) Solve. $(2-5)\times(3)=$

First subtract the numbers in brackets, $2-5=-3 \rightarrow (-3)\times(3)=$

Now use this formula: (negative) × (positive) = negative
$(-3)\times(3)=-9$

2) Solve. $(-12)+(48\div 6)=$

First divided 48 by 6 , the numbers in brackets, $48\div 6=8$

$=(-12)+(8)=-12+8=-4$

Find each product or quotient.

1) $(-7)\times(-8)=$

2) $(-4)\times 5=$

3) $5\times(-11)=$

4) $(-5)\times(-20)=$

5) $-(2)\times(-8)\times 3=$

6) $(12-4)\times(-10)=$

7) $16\div(-4)=$

8) $(-25)\div(-5)=$

9) $(-40)\div(-8)=$

10) $64\div(-8)=$

11) $(-49)\div 7=$

12) $(-112)\div(-4)=$

Order of Operations

Step-by-step guide:

When there is more than one math operation, use PEMDAS:

- ✓ Parentheses
- ✓ Exponents
- ✓ Multiplication and Division (from left to right)
- ✓ Addition and Subtraction (from left to right)

Examples:

1) Solve. $(5+7) \div (3^2 \div 3) =$

First simplify inside parentheses: $(12) \div (9 \div 3) = (12) \div (3) =$
Then: $(12) \div (3) = 4$

2) Solve. $(11 \times 5) - (12 - 7) =$

First simplify inside parentheses: $(11 \times 5) - (12 - 7) = (55) - (5) =$

Then: $(55) - (5) = 50$

Evaluate each expression.

1) $5 + (4 \times 2) =$
2) $13 - (2 \times 5) =$
3) $(16 \times 2) + 18 =$
4) $(12 - 5) - (4 \times 3) =$
5) $25 + (14 \div 2) =$
6) $(18 \times 5) \div 5 =$
7) $(48 \div 2) \times (-4) =$
8) $(7 \times 5) + (25 - 12) =$
9) $64 + (3 \times 2) + 8 =$
10) $(20 \times 5) \div (4 + 1) =$
11) $(-9) + (12 \times 6) + 15 =$
12) $(7 \times 8) - (56 \div 4) =$

Scientific Notation

Step-by-step guide:

- ✓ It is used to write very big or very small numbers in decimal form.
- ✓ In scientific notation all numbers are written in the form of:

$$m \times 10^n$$

Decimal notation	Scientific notation
5	5×10^0
$-25{,}000$	-2.5×10^4
0.5	5×10^{-1}
2,122.456	$2{,}122456 \times 10^3$

Examples:

1) *Write* $\mathbf{0.00012}$ *in scientific notation.*

First, move the decimal point to the right so that you have a number that is between 1 and 10. Then: $N = 1.2$

Second, determine how many places the decimal moved in step 1 by the power of 10.

Then: 10^{-4} → When the decimal moved to the right, the exponent is negative.

Then: $0.00012 = 1.2 \times 10^{-4}$

2) *Write* $\mathbf{8.3 \times 10^{-5}}$ *in standard notation.*

10^{-5} → When the decimal moved to the right, the exponent is negative.

Then: $8.3 \times 10^{-5} = 0.000083$

Write each number in scientific notation.

1) $0.000325 =$

2) $0.00023 =$

3) $56{,}000{,}000 =$

4) $21{,}000 =$

Write each number in standard notation.

5) $3 \times 10^{-1} =$

6) $5 \times 10^{-2} =$

7) $1.2 \times 10^3 =$

8) $2 \times 10^{-4} =$

Integers and Absolute Value

Step-by-step guide:

✓ To find an absolute value of a number, just find its distance from 0 on number line! For example, the distance of 12 and -12 from zero on number line is 12!

Examples:

1) Solve. $\frac{|-18|}{9} \times |5-8| =$

First find $|-18|$, →the absolute value of -18 is 18, then: $|-18| = 18$

$\frac{18}{9} \times |5-8| =$

Next, solve $|5-8|$, → $|5-8| = |-3|$, the absolute value of -3 is 3. $|-3| = 3$

Then: $\frac{18}{9} \times 3 = 2 \times 3 = 6$

2) Solve. $|10-5| \times \frac{|-2\times6|}{3} =$

First solve $|10-5|$, →$|10-5| = |5|$, the absolute value of 5 is 5, $|5| = 5$

$5 \times \frac{|-2\times6|}{3} =$

Now solve $|-2\times6|$, → $|-2\times6| = |-12|$, the absolute value of -12 is 12, $|-12| = 12$

Then: $5 \times \frac{12}{3} = 5 \times 4 = 20$

✍ ***Evaluate the value.***

1) $8 - |2-14| - |-2| =$

2) $|-2| - \frac{|-10|}{2} =$

3) $\frac{|-36|}{6} \times |-6| =$

4) $\frac{|5\times-3|}{5} \times \frac{|-20|}{4} =$

5) $|2\times-4| + \frac{|-40|}{5} =$

6) $\frac{|-28|}{4} \times \frac{|-55|}{11} =$

7) $|-12+4| \times \frac{|-4\times5|}{2}$

8) $\frac{|-10\times3|}{2} \times |-12| =$

Variable Expressions

Step-by-step guide:

- ✓ In algebra, a variable is a letter used to stand for a number. The most common letters are: $x, y, z, a, b, c, m, and\ n$.
- ✓ algebraic expression is an expression contains integers, variables, and the math operations such as addition, subtraction, multiplication, division, etc.
- ✓ In an expression, we can combine "like" terms. (values with same variable and same power)

Examples:

1) Simplify this expression. $(10x + 2x + 3) =?$
   Combine like terms. Then: $(10x + 2x + 3) = 12x + 3$ (remember you cannot combine variables and numbers.
2) Simplify this expression. $12 - 3x^2 + 9x + 5x^2 =?$
   Combine "like" terms: $-3x^2 + 5x^2 = 2x^2$
   Then: $12 - 3x^2 + 9x + 5x^2 = 12 + 2x^2 + 9x$. Write in standard form (biggest powers first): $2x^2 + 9x + 12$

Simplify each expression.

1) $(2x + x + 3 + 24) =$
2) $(-28x - 20x + 24) =$
3) $7x + 3 - 3x =$
4) $-2 - x^2 - 6x^2 =$
5) $3 + 10x^2 + 2 =$
6) $8x^2 + 6x + 7x^2 =$
7) $5x^2 - 12x^2 + 8x =$
8) $2x^2 - 2x - x =$
9) $4x + (12 - 30x) =$
10) $10x + (80x - 48) =$
11) $(-18x - 54) - 5 =$
12) $2x^2 + (-8x)$

Simplifying Polynomial Expressions

Step-by-step guide:

- ✓ In mathematics, a polynomial is an expression consisting of variables and coefficients that involves only the operations of addition, subtraction, multiplication, and non-negative integer exponents of variables.

$$P(x) = a_n x^n + a_{n-1}x^{n-1} + \ldots + a_2x^2 + a_1x + a_0$$

Examples:

1) Simplify this Polynomial Expressions. $4x^2 - 5x^3 + 15x^4 - 12x^3 =$
   Combine "like" terms: $-5x^3 - 12x^3 = -17x^3$
   Then: $4x^2 - 5x^3 + 15x^4 - 12x^3 = 4x^2 - 17x^3 + 15x^4$
   Then write in standard form: $4x^2 - 17x^3 + 15x^4 = 15x^4 - 17x^3 + 4x^2$

2) Simplify this expression. $(2x^2 - x^4) - (4x^4 - x^2) =$
   First use distributive property: → multiply $(-)$ into $(4x^4 - x^2)$
   $(2x^2 - x^4) - (4x^4 - x^2) = 2x^2 - x^4 - 4x^4 + x^2$
   Then combine "like" terms: $2x^2 - x^4 - 4x^4 + x^2 = 3x^2 - 5x^4$
   And write in standard form: $3x^2 - 5x^4 = -5x^4 + 3x^2$

Simplify each polynomial.

1) $(2x^3 + 5x^2) - (12x + 2x^2) =$ ______________________

2) $(2x^5 + 2x^3) - (7x^3 + 6x^2) =$ ______________________

3) $(12x^4 + 4x^2) - (2x^2 - 6x^4) =$ ______________________

4) $14x - 3x^2 - 2(6x^2 + 6x^3) =$ ______________________

5) $(5x^3 - 3) + 5(2x^2 - 3x^3) =$ ______________________

6) $(4x^3 - 2x) - 2(4x^3 - 2x^4) =$ ______________________

7) $2(4x - 3x^3) - 3(3x^3 + 4x^2) =$ ______________________

8) $(2x^2 - 2x) - (2x^3 + 5x^2) =$ ______________________

Translate Phrases into an Algebraic Statement

Step-by-step guide:

Translating key words and phrases into algebraic expressions:

- ✓ Addition: plus, more than, the sum of, etc.
- ✓ Subtraction: minus, less than, decreased, etc.
- ✓ Multiplication: times, product, multiplied, etc.
- ✓ Division: quotient, divided, ratio, etc.

Examples:

Write an algebraic expression for each phrase.

1) Eight more than a number is 20.
More than mean plus a number $= x$
Then: $8 + x = 20$

2) 5 times the sum of 8 and x.
Sum of 8 and x: $8 + x$. Times means multiplication. Then: $5 \times (8 + x)$

Write an algebraic expression for each phrase.

1) 4 multiplied by x. ________________

2) Subtract 8 from y. ________________

3) 6 divided by x. ________________

4) 12 decreased by y. ________________

5) Add y to 9. ________________

6) The square of 5. ________________

7) x raised to the fourth power. ________________

8) The sum of nine and a number. ________________

9) The difference between sixty–four and y. ________________

10) The quotient of twelve and a number. ________________

11) The quotient of the square of x and 7. ________________

The difference between x and 8 is 22. ________________

Answers of Worksheets – Chapter 1

Multiplying and Dividing Integers

1) 56
2) -20
3) -55
4) 100
5) 48
6) -80
7) -4
8) 5
9) 5
10) -8
11) -7
12) 28

Order of Operations

1) 13
2) 3
3) 50
4) -5
5) 32
6) 18
7) -96
8) 48
9) 78
10) 20
11) 78
12) 42

Scientific Notation

1) 3.25×10^{-4}
2) 2.3×10^{-4}
3) 5.6×10^{7}
4) 2.1×10^{4}
5) 0.3
6) 0.05
7) 1,200
8) 0.0002

Integers and Absolute Value

1) -6
2) -3
3) 36
4) 15
5) 16
6) 35
7) 80
8) 180

Simplifying Variable Expressions

1) $3x + 27$
2) $-48x + 24$
3) $4x + 3$
4) $-7x^2 - 2$
5) $10x^2 + 5$
6) $15x^2 + 6x$
7) $-7x^2 + 8x$
8) $2x^2 - 3x$
9) $-26x + 12$
10) $90x - 48$

11) $-18x - 59$

12) $2x^2 - 8x$

Simplifying Polynomial Expressions

1) $2x^3 + 3x^2 - 12x$
2) $2x^5 - 5x^3 - 6x^2$
3) $18x^4 + 2x^2$
4) $-12x^3 - 15x^2 + 14x$
5) $-10x^3 + 10x^2 - 3$
6) $4x^4 - 4x^3 - 2$
7) $-15x^3 - 12x^2 + 8x$
8) $-2x^3 - 3x^2 - 2x$

Translate Phrases into an Algebraic Statement

1) $4x$
2) $y - 8$
3) $\frac{6}{x}$
4) $12 - y$
5) $y + 9$
6) 5^2
7) x^4
8) $9 + x$
9) $64 - y$
10) $\frac{12}{x}$
11) $\frac{x^2}{7}$
12) $x - 8 = 22$

Chapter 2: Equations and Inequalities

Topics that you'll learn in this chapter:

- ✓ Solving Multi– Step Equations
- ✓ Slope and Intercepts
- ✓ Graphing Linear Inequalities
- ✓ Solving Compound Inequalities
- ✓ Solving Absolute Value Equations
- ✓ Solving Absolute Value Inequalities
- ✓ Graphing Absolute Value Inequalities

Solving Multi–Step Equations

Step-by-step guide:

- ✓ Combine "like" terms on one side.
- ✓ Bring variables to one side by adding or subtracting.
- ✓ Simplify using the inverse of addition or subtraction.
- ✓ Simplify further by using the inverse of multiplication or division.

Examples:

1) *Solve this equation.* $-(2-x)=5$

First use Distributive Property: $-(2-x)=-2+x$

Now solve by adding 2 to both sides of the equation. $-2+x=5 \rightarrow -2+x+2=5+2$

Now simplify: $-2+x+2=5+2 \rightarrow x=7$

2) *Solve this equation.* $4x+10=25-x$

First bring variables to one side by adding x to both sides.

$4x+10+x=25-x+x \rightarrow 5x+10=25$. Now, subtract 10 from both sides:

$5x+10-10=25-10 \rightarrow 5x=15$

Now, divide both sides by 5: $5x=15 \rightarrow 5x \div 5=\frac{15}{5} \rightarrow x=3$

✍ ***Solve each equation.***

1) $-3(2+x)=3$
2) $-2(4+x)=4$
3) $20=-(x-8)$
4) $2(2-2x)=20$
5) $-12=-(2x+8)$
6) $5(2+x)=5$
7) $2(x-14)=4$
8) $-28=2x+12x$
9) $3x+15=-x-5$
10) $2(3+2x)=-18$
11) $12-2x=-8-x$
12) $10-3x=14+x$

Slope and Intercepts

Step-by-step guide:

- ✓ The slope of a line represents the direction of a line on the coordinate plane.
- ✓ A coordinate plane contains two perpendicular number lines. The horizontal line is x and the vertical line is y. The point at which the two axes intersect is called the origin. An ordered pair (x, y) shows the location of a point.
- ✓ A line on coordinate plane can be drawn by connecting two points.
- ✓ To find the slope of a line, we need two points.
- ✓ The slope of a line with two points A (x_1, y_1) and B (x_2, y_2) can be found by using this formula: $\frac{y_2 - y_1}{x_2 - x_1} = \frac{rise}{\text{run}}$

Examples:

1) *Find the slope of the line through these two points:* $(2, -10)$ *and* $(3, 6)$.

Slope $= \frac{y_2 - y_1}{x_2 - x_1}$. Let (x_1, y_1) be $(2, -10)$ and (x_2, y_2) be $(3, 6)$. *Then:* slope $= \frac{y_2 - y_1}{x_2 - x_1} = \frac{6-(-10)}{3-2} = \frac{6+10}{1} = \frac{16}{1} = 16$

2) *Find the slope of the line containing two points* $(8, 3)$ and $(-4, 9)$.

Slope $= \frac{y_2 - y_1}{x_2 - x_1} \rightarrow (x_1, y_1) = (8,3)$ and $(x_2, y_2) = (-4, 9)$. *Then:* slope $= \frac{y_2 - y_1}{x_2 - x_1} = \frac{9-3}{-4-8} = \frac{6}{-12} = \frac{1}{-2} = -\frac{1}{2}$

Find the slope of the line through each pair of points.

1) $(1, 1), (2, 3)$
2) $(-1, 2), (0, 3)$
3) $(3, -1), (2, 3)$
4) $(-2, -1), (0, 5)$
5) $(5, 1), (2, 4)$
6) $(-3,1), (-2, 4)$
7) $(6, 2), (7, 4)$
8) $(6, -5), (3, 4)$
9) $(12, -9), (11, -8)$
10) $(7, 4), (5, -2)$
11) $(1, 1), (3, 5)$
12) $(7, -12), (5, 10)$

Solving Inequalities

Step-by-step guide:

- ✓ Similar to equations, first isolate the variable by using inverse operation.
- ✓ For dividing or multiplying both sides by negative numbers, flip the direction of the inequality sign.

Examples:

1) *Solve and graph the inequality.* $x + 2 \geq 3$.

Subtract 2 from both sides. $x + 2 \geq 3 \rightarrow x + 2 - 2 \geq 3 - 2$, then: $x \geq 1$

2) *Solve this inequality.* $x - 1 \leq 2$

Add 1 to both sides. $x - 1 \leq 2 \rightarrow x - 1 + 1 \leq 2 + 1$, then: $x \leq 3$

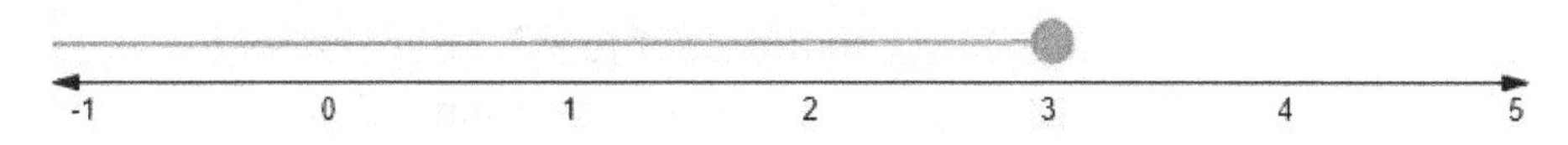

Solve each inequality and graph it.

1) $2x \geq 12$

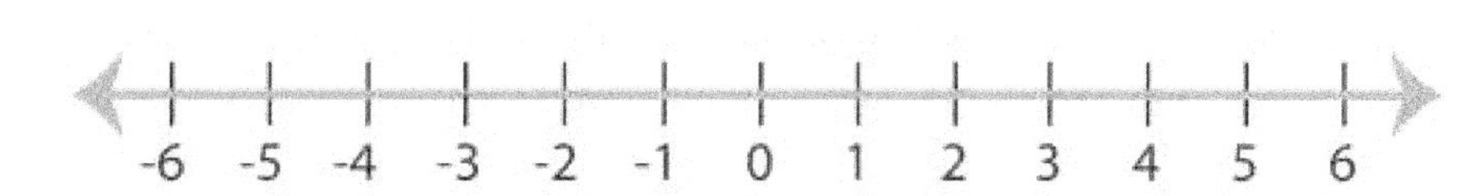

2) $4 + x \leq 5$

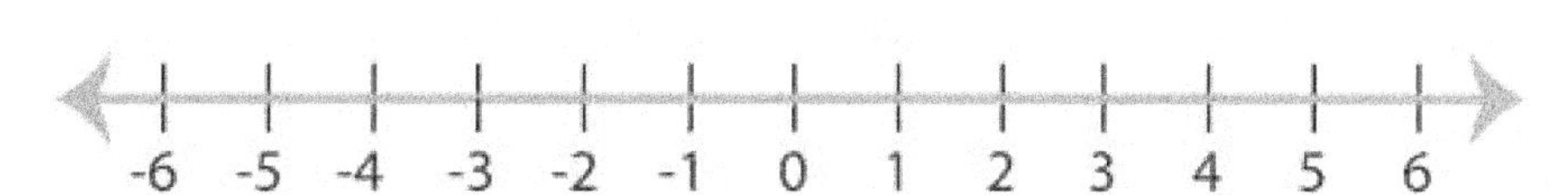

3) $x + 3 \leq -3$

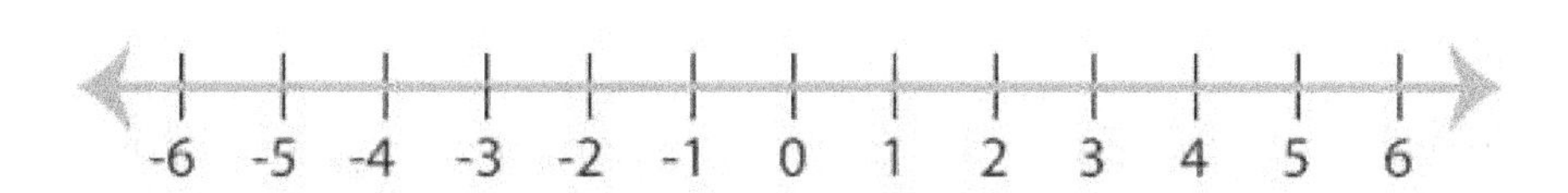

4) $4x \geq 16$

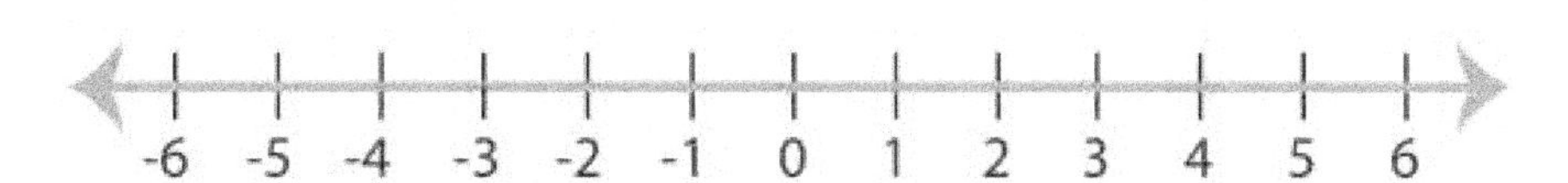

5) $9x \leq 18$

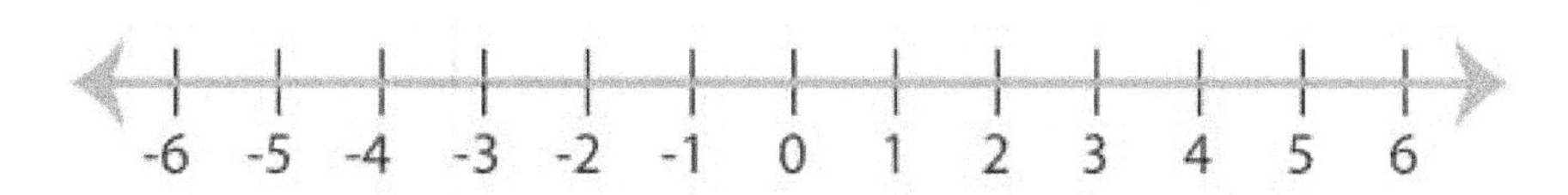

Graphing Linear Inequalities

Step-by-step guide:

- ✓ First, graph the "equals" line.
- ✓ Choose a testing point. (it can be any point on both sides of the line.)
- ✓ Put the value of (x, y) of that point in the inequality. If that works, that part of the line is the solution. If the values don't work, then the other part of the line is the solution.

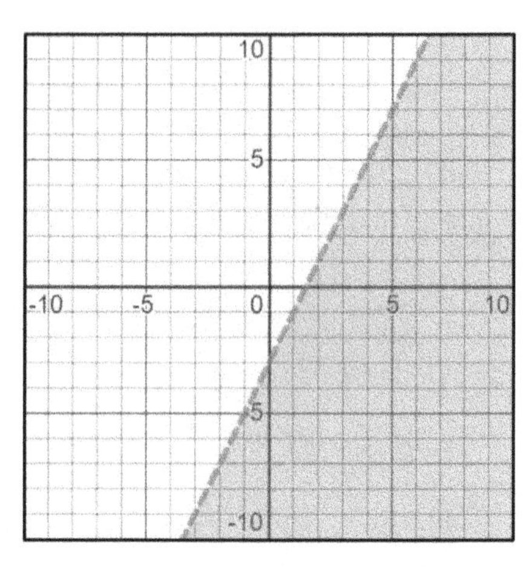

Example:

Sketch the graph of $y < 2x - 3$. First, graph the line:

$y = 2x - 3$. The slope is 2 and y-intercept is -3. Then, choose a testing point. The easiest point to test is the origin: $(0, 0)$

$(0,0) \rightarrow y < 2x - 3 \rightarrow 0 < 2(0) - 3 \rightarrow 0 < -3$

0 is not less than -3. So, the other part of the line (on the right side) is the solution.

Sketch the graph of each linear inequality.

1) $y > 3x - 1$ **2)** $y < -x + 4$ **3)** $y \leq -5x + 8$

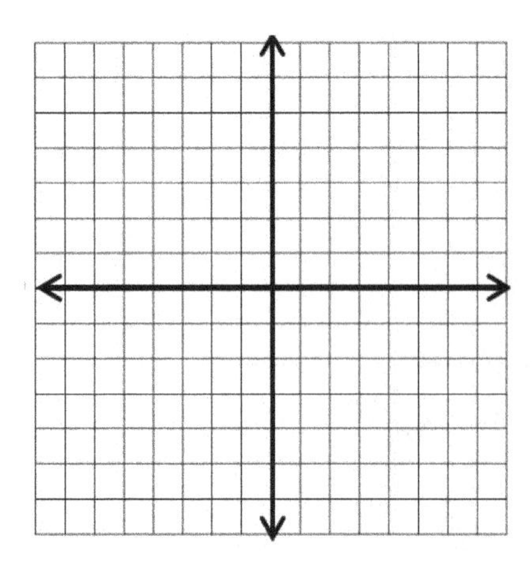

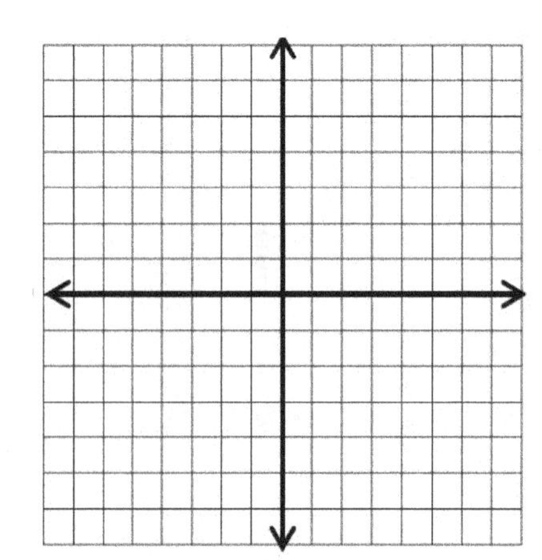

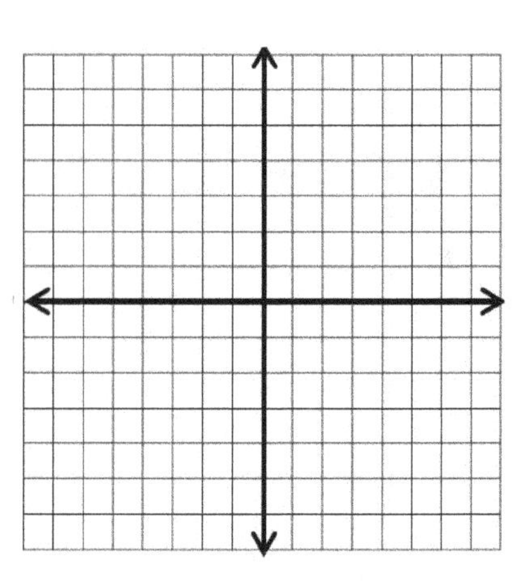

Solving Compound Inequalities

Step-by-step guide:

- ✓ Isolate the variable.
- ✓ Simplify using the inverse of addition or subtraction.
- ✓ Simplify further by using the inverse of multiplication or division.

Examples:

1) *Solve this inequality.* $2x - 2 \le 6$

First add 2 to both sides: $2x - 2 + 2 \le 6 + 2 \rightarrow 2x \le 8$,

Now, divide both sides by 2: $2x \le 8 \rightarrow x \le 4$

2) *Solve this inequality.* $2x - 4 \le 8$

First add 4 to both sides: $2x - 4 + 4 \le 8 + 4$

Then simplify: $2x - 4 + 4 \le 8 + 4 \rightarrow 2x \le 12$,

Now divide both sides by 2: $\frac{2x}{2} \le \frac{12}{2} \rightarrow x \le 6$

Solve each inequality.

1) $2x - 8 \le 6$

2) $8x - 2 \le 14$

3) $-5 + 3x \le 10$

4) $2(x - 3) \le 6$

5) $7x - 5 \le 9$

6) $4x - 21 < 19$

7) $2x - 3 < 21$

8) $17 - 3x \ge -13$

9) $3 + 2x \ge 19$

10) $3 + 2x \ge 19$

Solving Absolute Value Equations

Step-by-step guide:

- ✓ Isolate the absolute value.
- ✓ Take off the absolute value sign and solve the equation.
- ✓ Write another equation with negative sign of the answer of the absolute value equation.
- ✓ Plugin the value of the variable into the original equation and check your answer.

Examples:

1) *Solve this equation.* $|x+1|=2$
   First $+1=2$, $x+1=-2$
   $x=2-1=1$ or $x=-2-1=-3$
2) *Solve this equation.* $|x+2|=5$
   First $+2=5$, $x+2=-5$
   $x=5-2=3$ or $x=-5-2=-7$

Solve each equation

1) $|a|+12=22$
2) $3|x+4|=45$
3) $-2\,|x|=-24$
4) $|5+x|=5$
5) $|-4x|=16$
6) $-2|x+2|=-12$
7) $|6x|+4=70$
8) $|-5\,m|=30$
9) $|-4+5x|=16$
10) $|\frac{x}{4}|=5$

Solving Absolute Value Inequalities

Step-by-step guide:

- ✓ An Absolute value can never be negative. Therefore, absolute value cannot be less than a negative number.

- ✓ To solve an absolute value inequality, first isolate it on one side of the inequality. Then, if the inequality sign is greater (or greater and equal), write the value inside absolute value greater than the value provided and less than the negative of the value provided and solve. If the sign is less than, then do the opposite.

Examples:

1) *Solve this equation.* $|2x| \geq 24$

First $2x \geq 24$ and $2x \leq -24$

$x \geq 12$ or $x \leq -12$

2) *Solve this equation.* $|3x| \geq 36$

First $3x \geq 36$ and $3x \leq -36$

$x \geq 12$ or $x \leq -12$

Solve each equation

1) $|x| + 4 \geq 6$
2) $|x - 2| - 6 < 5$
3) $3 + |2 + x| < 5$
4) $|x + 7| - 9 < -6$
5) $|x| - 3 > 2$
6) $|x| - 2 > 0$
7) $|3x| \leq 15$
8) $|x + 4| \leq 8$
9) $|3x| \leq 24$
10) $|x - 8| - 10 < 26$

Graphing Absolute Value Inequalities

Step-by-step guide:

- ✓ Solve the absolute value inequality using the Properties of Inequality.
- ✓ Find two key values.
- ✓ Represent absolute value inequalities on a number line.

Examples:

1) *Solve and graph this equation* $|-8x| < 32$:
Use absolute rule: if $|u| < a, a > 0$ *then:* $-a < u < a$

Then: $-32 < -8x < 32$, $-8x > -32$ *and* $-8x < 32$

$-8x > -32 \rightarrow x < 4$ and $-8x < 32 \rightarrow x > -4$

Then: $-4 < x < 4$

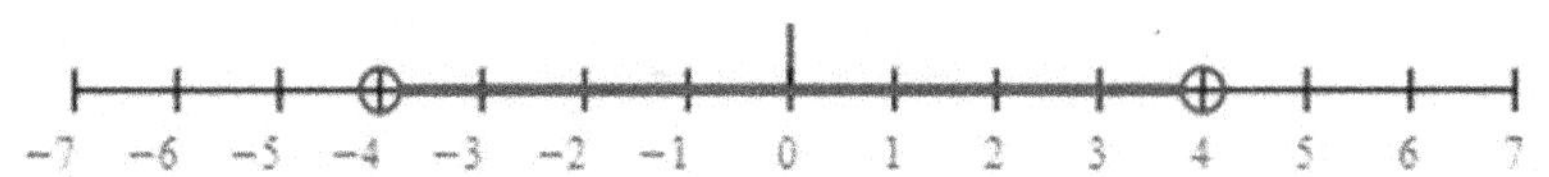

2) *Solve and graph this equation* $|10 + 4x| < 14$:
Use absolute rule: if $|u| < a, a > 0$ *then:* $-a < u < a$
First $-14 < 10 + 4x < 14$, $10 + 4x > -14$ and $10 + 4x < 14$.
$10 + 4x > -14 \rightarrow x > -6$
$10 + 14x < 14 \rightarrow x < 1$
Then: $-6 < x < 1$

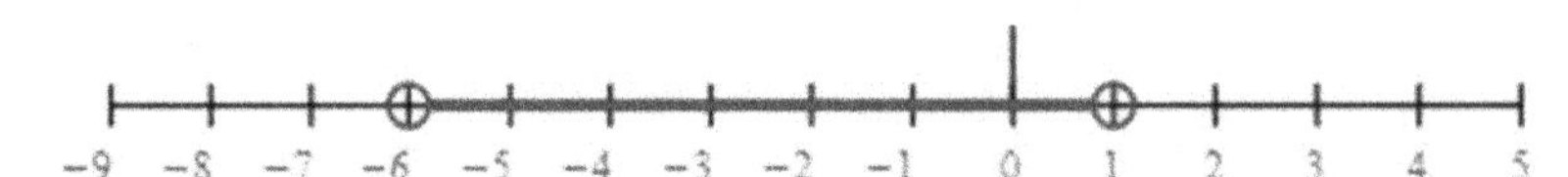

✍ *Solve each inequality and graph its solution*

1) $|3 - 9x| \leq 60$

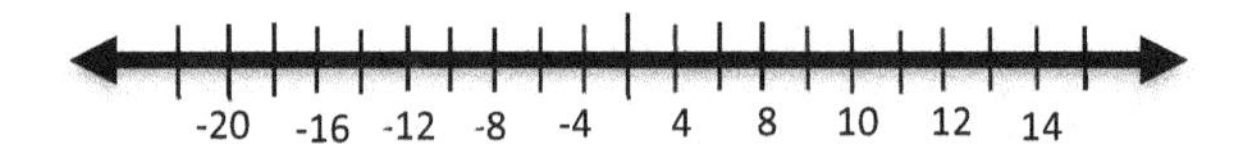

2) $|7x + 4| \geq 74$

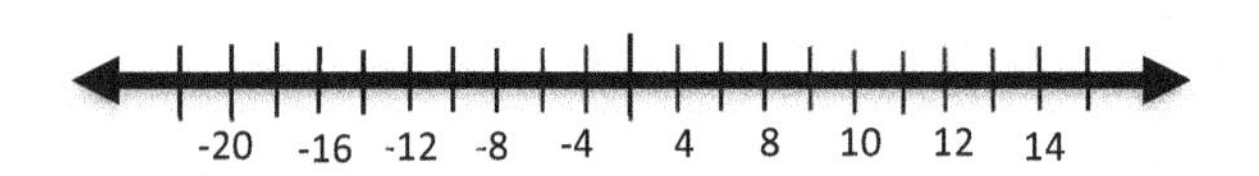

3) $|x| - 6 \leq -1$

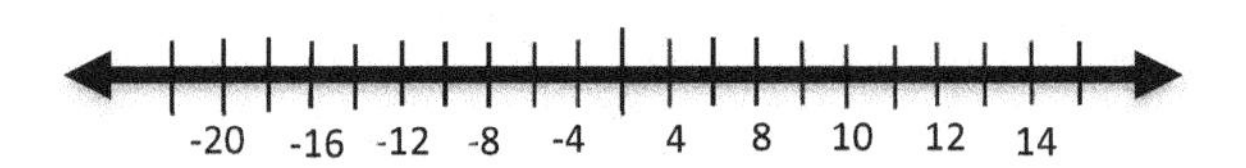

Answers of Worksheets – Chapter 2

Multi–Step Equations

1) -3
2) -6
3) -12
4) -4
5) 2
6) -1
7) 16
8) -2
9) -5
10) -6
11) 20
12) -1

Slope and intercepts

1) 2
2) 1
3) -4
4) 3
5) -1
6) 3
7) 2
8) -3
9) -1
10) 3
11) 2
12) -11

Solving Inequities

1) 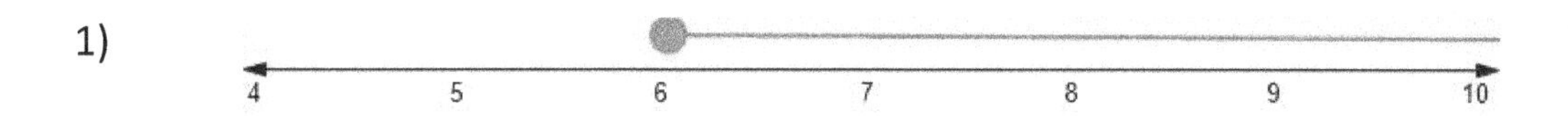

2) 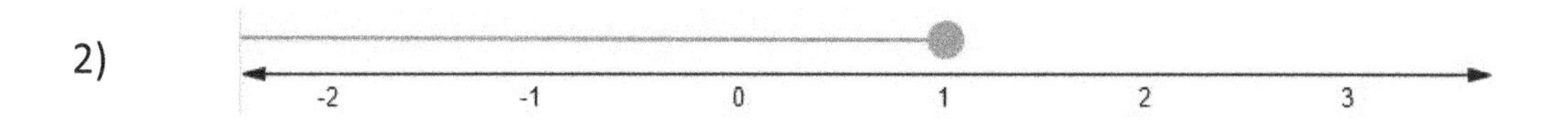

3) 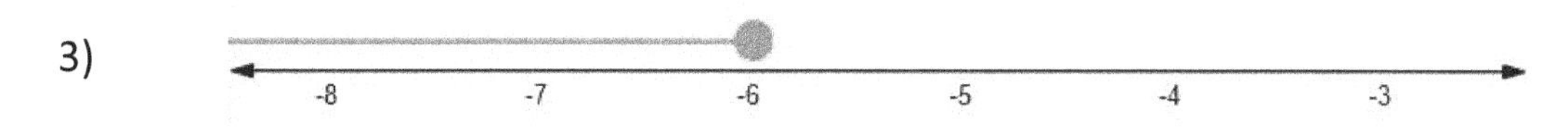

4) 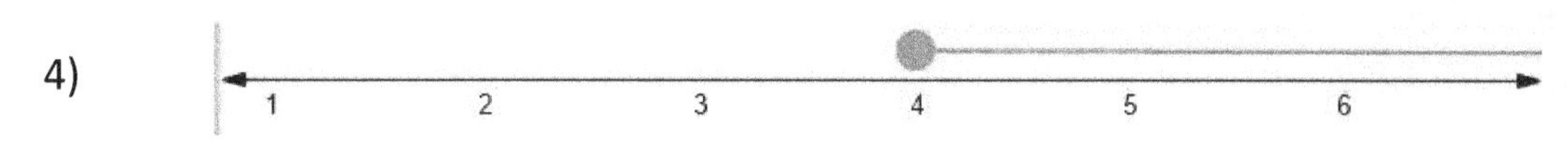

5) 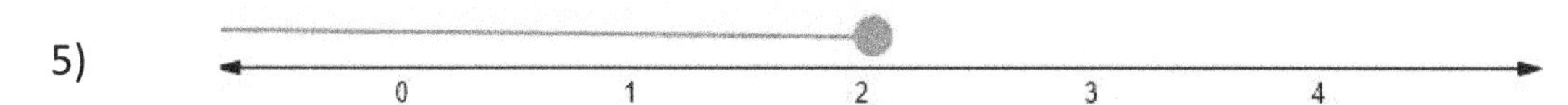

Graphing Linear Inequalities

1) $y > 3x - 1$

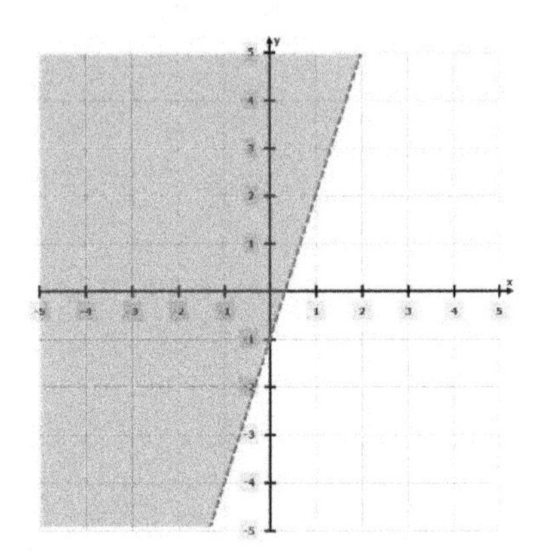

2) $y < -x + 4$

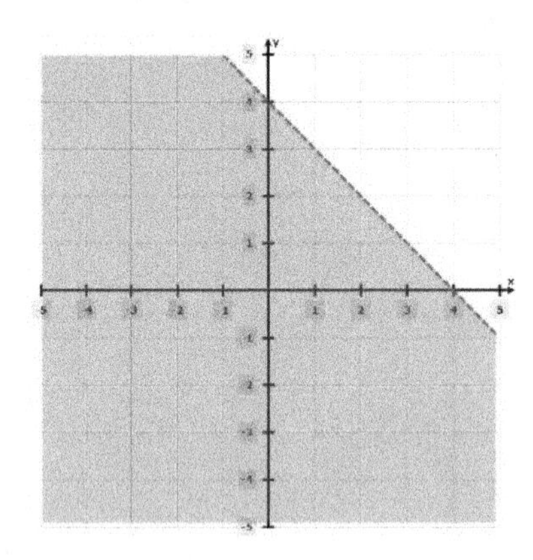

3) $y \leq -5x + 8$

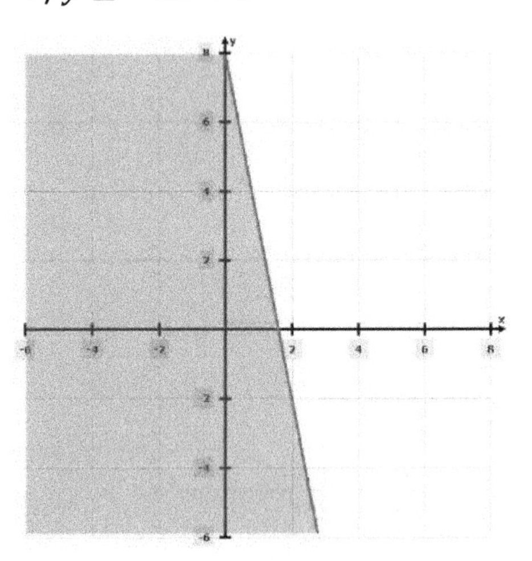

Solving Compound Inequalities

1) $x \leq 7$
2) $x \leq 2$
3) $x \leq 5$
4) $x \leq 6$
5) $x \leq 2$
6) $x < 10$
7) $x < 12$
8) $x \geq 10 \ \ x \leq 10$
9) $x < 3$
10) $x \geq 8$

Solving Absolute Value Equations

1) $10, -10\}$
2) $\{11, -19\}$
3) $\{12, -12\}$
4) $\{0, -10\}$
5) $\{-4, 4\}$
6) $\{4, -8\}$
7) $\{11, -11\}$
8) $\{-6, 6\}$
9) $\{4, -\frac{12}{5}\}$
10) $\{20, 20\}$

Solving Absolute Value Inequalities

1) $x \geq 3$ *or* $x \leq -3$
2) $x < 13$ *and* $x > -9$
3) $x < 0$ *and* $x > -4$
4) $x < -4$ *and* $x > -10$
5) $x > 5$ *or* $x < -5$
6) $x > 2$ *or* $x < -2$
7) $x \leq 5$ *and* $x \geq -5$
8) $-12 \leq x \leq 4$
9) $x \leq 8$ *and* $x \geq -8$
10) $4 < x < 12$

Graphing Absolute Value Inequalities

1)

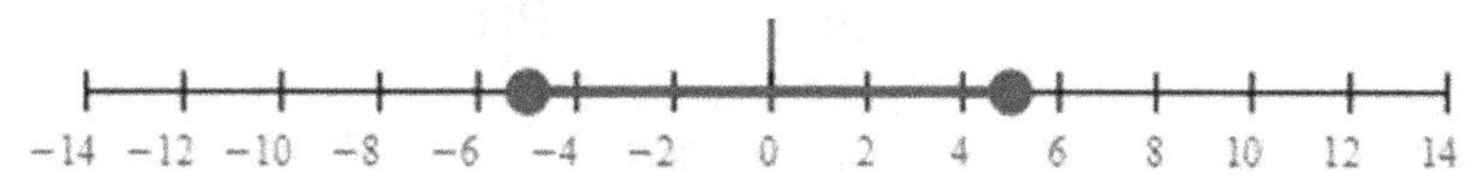

2)

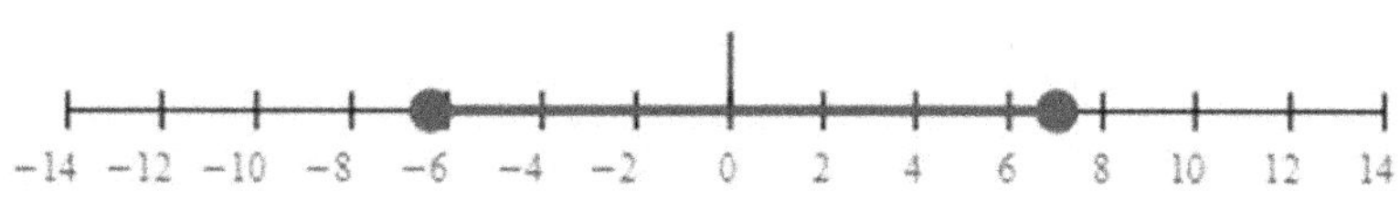

3)

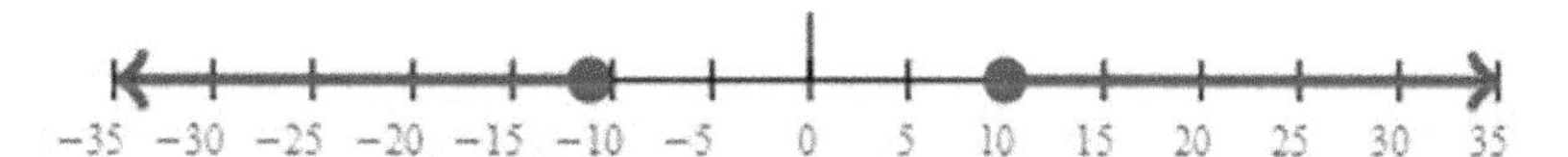

Chapter 3: System of Equations and Inequalities

Topics that you'll learn in this chapter:

- ✓ Systems of Two Equations
- ✓ Systems of Two Equations Word Problems
- ✓ Systems of 3 Variable Equations

Systems of Two Equations

Step-by-step guide:

- ✓ A system of equations contains two equations and two variables. For example, consider the system of equations: $x - y = 1, x + y = 5$
- ✓ The easiest way to solve a system of equation is using the elimination method. The elimination method uses the addition property of equality. You can add the same value to each side of an equation.
- ✓ For the first equation above, you can add $x + y$ to the left side and 5 to the right side of the first equation: $x - y + (x + y) = 1 + 5$. Now, if you simplify, you get: $x - y + (x + y) = 1 + 5 \rightarrow 2x = 6 \rightarrow x = 3$. Now, substitute 3 for the x in the first equation: $3 - y = 1$. By solving this equation, $y = 2$

Example:

What is the value of $x + y$ in this system of equations? $\begin{cases} 3x - 4y = -20 \\ -x + 2y = 10 \end{cases}$

Solving Systems of Equations by Elimination: $\begin{array}{r} 3x - 4y = -20 \\ \underline{-x + 2y = 10} \end{array}$ $\Rightarrow$ Multiply the second equation by 3, then add it to the first equation.

$\begin{array}{r} 3x - 4y = -20 \\ \underline{3(-x + 2y = 10)} \end{array} \Rightarrow \begin{array}{r} 3x - 4y = -20 \\ -3x + 6y = 30) \end{array} \Rightarrow 2y = 10 \Rightarrow y = 5$. Now, substitute 5 for y in the first equation and solve for x. $3x - 4(5) = -20 \rightarrow 3x - 20 = -20 \rightarrow x = 0$

✍ *Solve each system of equations.*

1) $-2x + 2y = 4$ $\quad x =$ ____
$-2x + y = 3$ $\quad y =$ ____

2) $-10x + 2y = -6$ $\quad x =$ ____
$6x - 16y = 48$ $\quad y =$ ____

3) $y = -8$ $\quad x =$ ____
$16x - 12y = 32$

4) $2y = -6x + 10$ $\quad x =$ ____
$10x - 8y = -6$ $\quad y =$ ____

5) $10x - 9y = -13$ $\quad x =$ ____
$-5x + 3y = 11$ $\quad y =$ ____

6) $-3x - 4y = 5$ $\quad x =$ ____
$x - 2y = 5$ $\quad y =$ ____

Systems of Two Equations Word Problems

Step-by-step guide:

- ✓ Define your variables, write two equations, and use elimination method for solving systems of equations.

Example:

Tickets to a movie cost $8 for adults and $5 for students. A group of friends purchased **20** tickets for $**115.00**. How many adults ticket did they buy? ____

Let x be the number of adult tickets and y be the number of student tickets. There are 20 tickets. Then: $x + y = 20$. The cost of adults' tickets is $8 and for students it is $5, and the total cost is $115. So, $8x + 5y = 20$. Now, we have a system of equations: $\begin{cases} x + y = 20 \\ 8x + 5y = 115 \end{cases}$

Multiply the first equation by -5 and add to the second equation: $-5(x + y = 20) = -5x - 5y = -100$

$8x + 5y + (-5x - 5y) = 115 - 100 \rightarrow 3x = 15 \rightarrow x = 5 \rightarrow 5 + y = 20 \rightarrow y = 15$. There are 5 adult tickets and 15 student tickets.

Solve each word problem.

1) Tickets to a movie cost $5 for adults and $3 for students. A group of friends purchased **18** tickets for $**82.00**. How many adults ticket did they buy? _____

2) At a store, Eva bought two shirts and five hats for $**154.00**. Nicole bought three same shirts and four same hats for $**168.00**. What is the price of each shirt? ______

3) A farmhouse shelters **10** animals, some are pigs, and some are ducks. Altogether there are **36** legs. How many pigs are there? ______

4) A class of **195** students went on a field trip. They took **19** vehicles, some cars and some buses. If each car holds **5** students and each bus hold **25** students, how many buses did they take? ______

Systems of 3 Variable Equations

Step-by-step guide:

- ✓ It's similar to systems of 2 equations.
- ✓ To solve systems of equations you can use elimination method or the substitution method.
- ✓ In elimination method, you need to remove one variable by adding or subtracting two equations or multiplying the equations by real numbers.
- ✓ For substitution method, solve for one variable in the first equation and plugin the result into the second and third equations. Then, you will have systems of two equations.

Examples: Solve system of equation. $\begin{cases} x - y - 2z = -6 \\ 3x + 2y = -25 \\ -4x + y - z = 12 \end{cases}$

Isolate x for $3x + 2y = -25$, then: $x = \frac{-25-2y}{3}$, Substitute $x = \frac{-25-2y}{3}$

$\begin{cases} \frac{-25-2y}{3} - y - 2z = -6 \\ -4(\frac{-25-2y}{3}) + y - z = 12 \end{cases}$, Solve: $\frac{-25-2y}{3} - y - 2z = -6 : y = -\frac{6z+7}{5}$

Now substitute $y = -\frac{6z+7}{5}$, Isolate z for $-4\left(\frac{-25-2\left(-\frac{6z+7}{5}\right)}{3}\right) - \frac{6z+7}{5} - z = 12 : z = 3$

Solve $y = -\frac{6z+7}{5}$ with $z = 3$, then: $y = -\frac{6(3)+7}{5} = -5$

Solve for $x = \frac{-25-2y}{3}$ with $y = -5$, then: $x = \frac{-25-2(-5)}{3} = -5$

Then: $x = -5, y = -5$ and $z = 3$

Solve each system of equations.

1) $x = 3y - 3z + 8$ $\quad x =$ ____

$z = 4x + 5y - 14$ $\quad y =$ ____

$3y + 2z = 14$ $\quad z =$ ____

2) $6x - 6y = -12$ $\quad x =$ ____

$2z = -6x - 6y + 18$ $\quad y =$ ____

$-8x + 10y + 2z = 16$ $\quad z =$ ____

3) $4x - 8z = 40$ $\quad x =$ ____

$-6x + 2y - 8z = 40$ $\quad y =$ ____

$-8x + 4y + 6z = -30$ $\quad z =$ ____

4) $2x - 4y + 2z = -12$ $\quad x =$ ____

$2x + 10z = -24$ $\quad y =$ ____

$-2x + 12y + 8z = 6$ $\quad z =$ ____

Answers of Worksheets – Chapter 3

Systems of Equations

1) $x = -1, y = 1$
2) $x = 0, y = -3$
3) $x = -4$
4) $x = 1, y = 2$
5) $x = -4, y = -3$
6) $x = 1, y = -2$

Systems of Equations Word Problems

1) 14
2) \$32
3) 8
4) 5

Systems of 3 variable equations

1) $(2, 2, 4)$
2) $(1, 3, -3)$
3) $(0, 0, -5)$
4) $(3, 3, -3)$

Chapter 4: Quadratic Functions

Math Topics that you'll learn today:

- ✓ Solving Quadratic Equations
- ✓ Solve Quadratic Inequalities
- ✓ Graphing Quadratic Functions In Vertex Form
- ✓ Graphing Quadratic Inequalities

"If people do not believe that mathematics is simple, it is only because they do not realize how complicated life is." — John von Neumann

Solving Quadratic Equations

Step-by-step guide:

✓ For $ax^2 + bx + c = 0$, the values of x which are the solutions of the equation are given by:

$$x = \frac{-b \pm \sqrt{b^2 - 4ac}}{2a}$$

Examples:

1) *Solve equation.* $x^2 + x - 20 = 2x$

Subtract $2x$ from both sides: $x^2 + x - 20 - 2x = 2x - 2x \rightarrow x^2 - x - 20 = 0$

The solution are:

$x_1 = \frac{-b \pm \sqrt{b^2 - 4ac}}{2a} \rightarrow x_2 = \frac{-(-1) \pm \sqrt{(-1)^2 - 4.1(-20)}}{2.1} =$

$x_1 = \frac{-(-1) + \sqrt{(-1)^2 - 4.1(-20)}}{2.1} = 5$ and $x_2 = \frac{-(-1) - \sqrt{(-1)^2 - 4.1(-20)}}{2.1} = -4$

2) *Solve equation.* $x^2 + 8x = -15$

Add 15 to both sides: $x^2 + 8x + 15 = -15 + 15 \rightarrow x^2 + 8x + 15 = 0$

$x_{1,2} = \frac{-b \pm \sqrt{b^2 - 4ac}}{2a} \rightarrow x_{1,2} = \frac{-8 \pm \sqrt{8^2 - 4.1.15}}{2.1}$

$x_1 = \frac{-8 + \sqrt{8^2 - 4.1.15}}{2.1} = -3$ and $x_2 = \frac{-8 - \sqrt{8^2 - 4.1.15}}{2.1} = -5$

Solve each equation by factoring.

1) $7x^2 - 14x = -7$

2) $6x^2 - 18x - 18 = 6$

3) $2x^2 + 6x - 24 = 12$

4) $2x^2 - 22x + 38 = -10$

5) $(2x + 5)(4x + 3) = 0$

6) $(x + 2)(x - 7) = 0$

7) $(x + 3)(x + 5) = 0$

8) $(5x + 7)(x + 4) = 0$

Solving Quadratic Inequalities

Step-by-step guide:

- ✓ A quadratic inequality is one that can be written in one of the following standard forms:

  $ax^2 + bx + c > 0,\ ax^2 + bx + c < 0,\ ax^2 + bx + c \geq 0,\ ax^2 + bx + c \leq 0$

- ✓ Solving a quadratic inequality is like solving equations. We need to find the solutions.

Examples:

1) Solve quadratic inequality. $-x^2 - 5x + 6 > 0$

   Factor: $-x^2 - 5x + 6 > 0 \rightarrow -(x-1)(x+6) > 0$

   Multiply both sides by -1: $\left(-(x-1)(x+6)\right)(-1) > 0(-1) \rightarrow (x-1)(x+6) < 0$

   Then the solution could be $-6 < x < 1$ or $-6 > x$ and $x > 1$. Choose a value between -1 and 6 and check. Let's try 0. Then: $-0^2 - 5(0) + 6 > 0 \rightarrow 6 > 0$. This is true! So, the answer is: $-6 < x < 1$

2) Solve quadratic inequality. $x^2 - 3x - 10 \geq 0$

   Factor: $x^2 - 3x - 10 \geq 0 \rightarrow (x+2)(x-5) \geq 0$. -2 and 5 are the solutions. Now, the solution could be $-2 \leq x \leq 5$ or $-6 \geq x$ and $x \geq 1$. Let's choose zero to check:

   $0^2 - 3(0) - 10 \geq 0 \rightarrow -10 \geq 0$, which is not true. So, $-6 \geq x$ and $x \geq 1$

Solve each quadratic inequality.

1) $x^2 + 4x - 5 > 0$
2) $x^2 - 2x - 3 \geq 0$
3) $x^2 - 1 < 0$
4) $17x^2 + 15x - 2 \geq 0$
5) $4x^2 + 20x - 11 < 0$
6) $12x^2 + 10x - 12 > 0$
7) $18x^2 + 23x + 5 \leq 0$
8) $-9x^2 + 29x - 6 \geq 0$
9) $-8x^2 + 6x - 1 \leq 0$
10) $5x^2 - 15x + 10 < 0$
11) $3x^2 - 5x \geq 4x^2 + 6$
12) $x^2 > 5x + 6$

Graphing Quadratic Functions

Step-by-step guide:

- ✓ Quadratic functions in vertex form: $y = a(x - h)^2 + k$ where (h, k) is the vertex of the function. The axis of symmetry is $x = h$
- ✓ To graph a quadratic function, first find the vertex, then substitute some values for x and solve for y.

Example:

Sketch the graph of $\mathbf{y = 3(x + 1)^2 + 2}$.

The vertex of $3(x + 1)^2 + 2$ is $(-1,2)$. Substitute zero for x and solve for y. $y = 3(0 + 1)^2 + 2 = 5$. The y Intercept is $(0,5)$.

Now, you can simply graph the quadratic function.

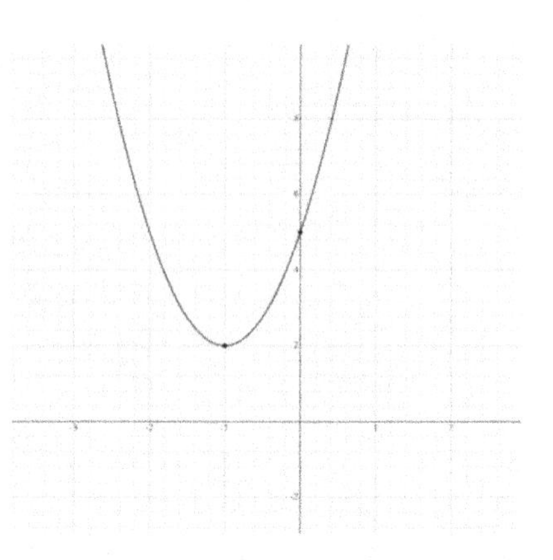

Sketch the graph of each function. Identify the vertex and axis of symmetry.

1) $y = 2(x - 3)^2 + 8$

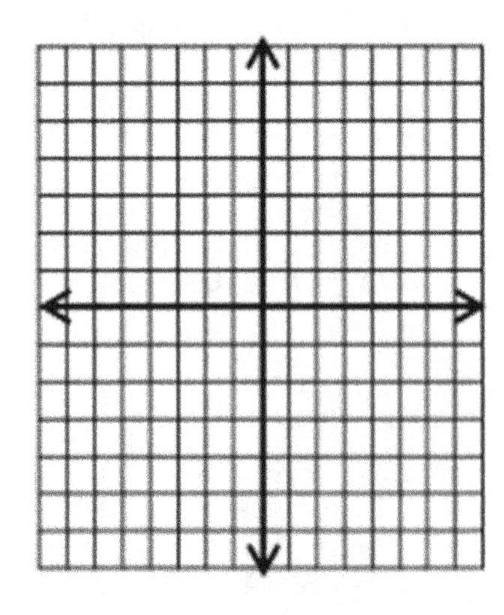

2) $y = x^2 - 8x + 19$

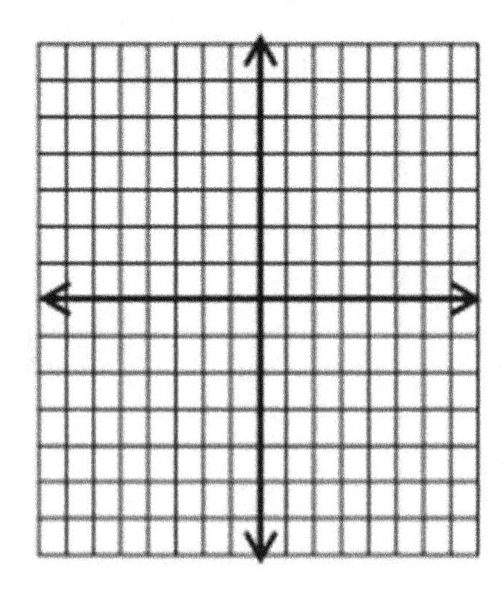

Graphing Quadratic inequalities

Step-by-step guide:

- ✓ A quadratic inequality is in the form $y > ax^2 + bx + c$ (or substitute $<, \leq,$ or $\geq$ for $>$).
- ✓ To graph a quadratic inequality, start by graphing the quadratic parabola. Then fill in the region either inside or outside of it, depending on the inequality.
- ✓ Choose a testing point and check the solution section.

Example: Sketch the graph of $y > 2x^2$.

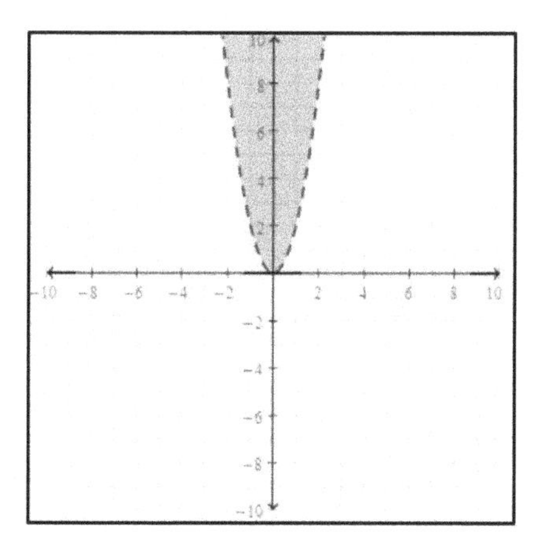

First, graph $y = 2x^2$

Since, the inequality sing is $>$, we need to use dash lines.

Now, choose a testing point inside the parabola. Let's choose $(0,2)$. $y > 2x^2 \rightarrow 2 > 2(0)^2 \rightarrow 2 > 0$

This is true. So, inside the parabola is the solution section.

Sketch the graph of each function.

1) $2y < -4x^2$

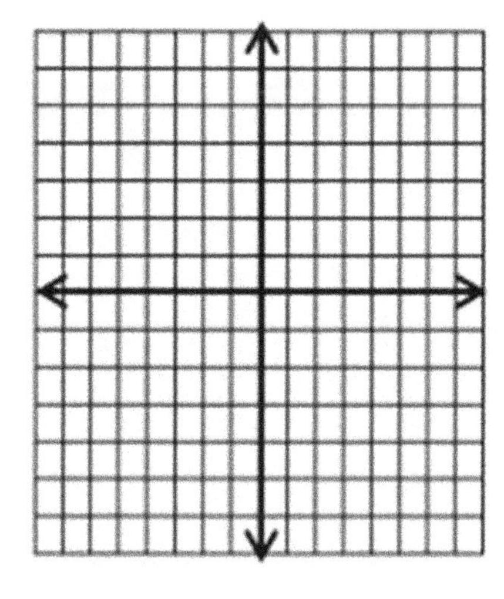

2) $y \geq 3x^2$

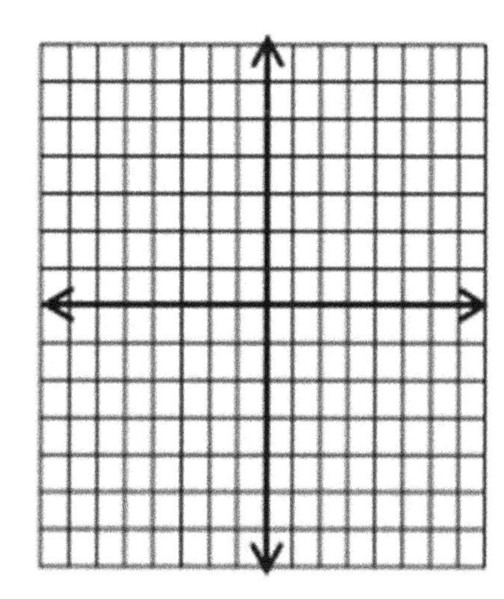

Answers of Worksheets – Chapter 4

Solving quadratic equations

1) $\{1\}$
2) $\{4, -1\}$
3) $\{3, -6\}$
4) $\{3, 8\}$
5) $\{-\frac{5}{2}, -\frac{3}{4}\}$
6) $\{-2, 7\}$
7) $\{-3, -5\}$
8) $\{-\frac{7}{5}, -4\}$

Solve quadratic inequalities

1) $x < -5$ or $x > 1$
2) $x \leq -1$ or $x \geq 3$
3) $-1 < x < 1$
4) $x \leq -1$ or $x \geq \frac{2}{17}$
5) $-\frac{11}{2} < x < \frac{1}{2}$
6) $x < -\frac{3}{2}$ or $x > \frac{2}{3}$
7) $-1 \leq x \leq -\frac{5}{18}$
8) $\frac{2}{9} \leq x \leq 3$
9) $x \leq \frac{1}{4}$ or $x \geq \frac{1}{2}$
10) $1 < x < 2$
11) $-3 \leq x \leq -2$
12) $x < -1$ or $x > 6$

Graphing quadratic functions

1) $x = 3$ asix of symetry

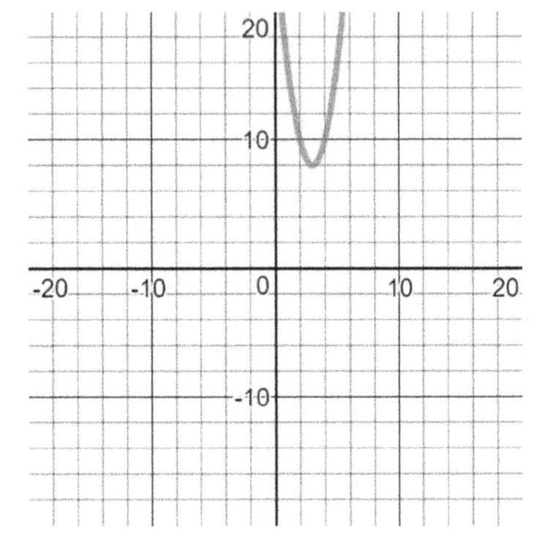

2) $x = 4$ asix of symetry

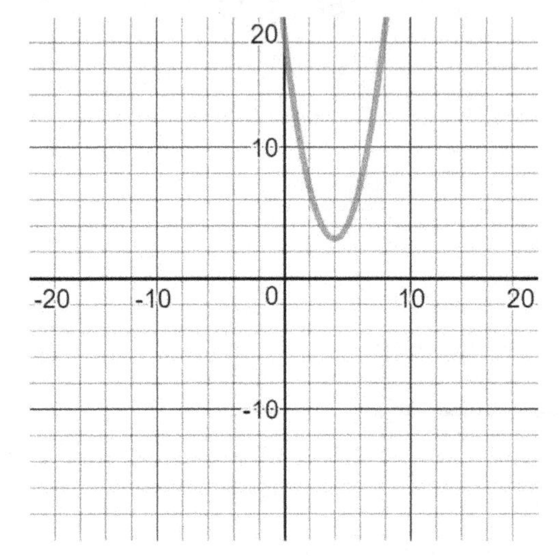

Graphing quadratic inequalities

1)

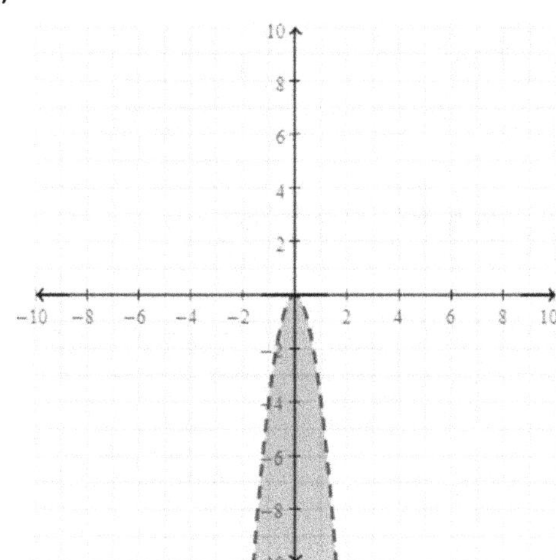

2)

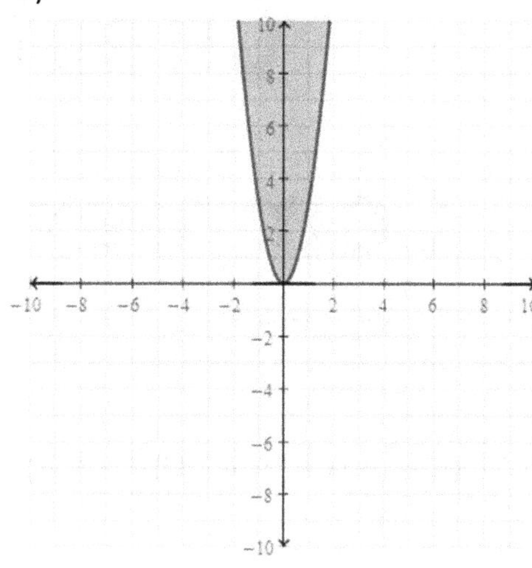

Chapter 5: Complex Numbers

Topics that you'll learn in this chapter:

- ✓ Adding and subtracting complex numbers
- ✓ Multiplying and dividing complex numbers
- ✓ Rationalizing imaginary denominators

Adding and Subtracting Complex Numbers

Step-by-step guide:

- ✓ A complex number is expressed in the form $a + bi$, where a and b are real numbers, and i, which is called an imaginary number, is a solution of the equation $x^2 = -1$
- ✓ For adding complex numbers: $(a + bi) + (c + di) = (a + c) + (b + d)i$
- ✓ For subtracting complex numbers: $(a + bi) - (c + di) = (a - c) + (b - d)i$

Examples:

1) Solve: $-8 + (2i) + (-8 + 6i)$

   Remove parentheses: $-8 + (2i) + (-8 + 6i) \rightarrow -8 + 2i - 8 + 6i$

   Group like terms: $-8 + 2i - 8 + 6i \rightarrow -8 - 8 + 2i + 6i$

   Add similar terms: $-8 - 8 + 2i + 6i = -16 + 8i$

2) Solve: $-2 + (-8 - 7i) - 9$

   Remove parentheses: $-2 + (-8 - 7i) - 9 \rightarrow -2 - 8 - 7i - 9$

   Combine like terms: $2 - 8 - 7i - 9 = -19 - 7i$

✍ ***Simplify.***

1) $(2 - 4i) + (-i) =$
2) $(-3i) + (3 + 5i) =$
3) $3 + (2 - 4i) =$
4) $(-5i) - (-5 + 2i) =$
5) $(5 + 3i) - (-4i) =$
6) $(8 + 5i) + (-7i) =$
7) $(9i) - (-6i + 10) =$
8) $(12i + 8) + (-7i) =$
9) $(13i) - (17 + 3i) =$
10) $(3 + 5i) + (8 + 3i) =$
11) $(8 - 3i) + (4 + i) =$
12) $(10 + 9i) + (6 + 8i) =$

Multiplying and Dividing Complex Numbers

Step-by-step guide:

- ✓ Multiplying complex numbers: $(a+bi)+(c+di)=(ac-bd)+(ad+bc)i$
- ✓ Dividing complex numbers: $\frac{a+bi}{c+di}=\frac{a+bi}{c+d}\times\frac{c-di}{c-di}=\frac{ac+b}{c^2-d^2}+\frac{bc+a}{c^2-d^2}i$
- ✓ Imaginary number rule: $i^2=-1$

Examples:

1) Solve: $(2-8i)(3-5i)$

   Use the rule: $(a+bi)+(c+di)=(ac-bd)+(ad+bc)i$

   $(2.3-(-8)9-5))+(2(-5)+(-8).3)i=-34-34i$

2) Solve: $\frac{2-3i}{2+i}=$

   Use the rule for dividing complex numbers:

$$\frac{a+bi}{c+di}=\frac{a+bi}{c+di}\times\frac{c-di}{c-di}=\frac{ac+bd}{c^2-d^2}+\frac{bc+ad}{c^2-d^2}i\rightarrow$$

$$\frac{2-3i}{2+i}\times\frac{2-i}{2-i}=\frac{(2\times(2)+(-3)(1)}{2^2-(-1)^2}+\frac{(-3\times(2)+(-1)(2)}{2^2-(-1)^2}i=\frac{1-8i}{5}=\frac{1}{5}-\frac{8}{5}i$$

Simplify.

1) $(5+4i)^2=$
2) $(4i)(-i)(2-5i)=$
3) $(2-8i)(3-5i)=$
4) $(-5+9i)(3+5i)=$
5) $(7+3i)(7+8i)=$
6) $2(3i)-(5i)(-8+5i)=$
7) $\frac{5}{-10i}=$
8) $\frac{4-3i}{-4i}=$
9) $\frac{5+9i}{i}=$
10) $\frac{12i}{-9+3i}=$
11) $\frac{-3-10i}{5i}=$
12) $\frac{9i}{3-i}=$

Rationalizing Imaginary Denominators

Step-by-step guide:

- ✓ Step 1: Find the conjugate (it's the denominator with different sign between the two terms.
- ✓ Step 2: Multiply numerator and denominator by the conjugate.
- ✓ Step 3: Simplify if needed.

Examples:

1) Solve: $\frac{5i}{2-3i}$

Multiply by the conjugate: $\frac{2+3i}{2+3i} \rightarrow \frac{5i(2+3i)}{(2-3i)(2+3i)} = \frac{-15+1}{(2-3i)(2+3i)}$

Use complex arithmetic rule: $(a+bi)(a-bi) = a^2 + b^2$

$(2-3i)(2+3i) = -2^2 + (-3)^2 = 13$, Then: $\frac{-15+}{(2-3i)(2+3i)} = \frac{-15+10i}{13}$

2) Solve: $\frac{4-9i}{-6i}$

Apply fraction rule: $\frac{4-9i}{-6i} = -\frac{4-9i}{6i}$

Multiply by the conjugate: $\frac{-i}{-i}$. $\quad -\frac{4-9i}{6i} = -\frac{(4-9i)(-i)}{6i(-i)} = -\frac{-9-4i}{6}$

Simplify.

1) $\frac{10-10i}{-5i} =$

2) $\frac{4-9i}{-6i} =$

3) $\frac{6+8i}{9i} =$

4) $\frac{8i}{-1+3i} =$

5) $\frac{5i}{-2-6i} =$

6) $\frac{-10-5}{-6+6i} =$

7) $\frac{-5-9i}{9+8i} =$

8) $\frac{-5-3i}{7-10i} =$

9) $\frac{-1+i}{-5i} =$

10) $\frac{-6-i}{i} =$

11) $\frac{-4-i}{9+5i} =$

12) $\frac{-3+i}{-2i} =$

Answers of Worksheets – Chapter 5

Adding and subtracting complex numbers

1) $2-5i$
2) $3+2i$
3) $5-4i$
4) $5-7i$
5) $5+7i$
6) $8-2i$
7) $10+15i$
8) $8+5i$
9) $-17+10i$
10) $11+8i$
11) $12-2i$
12) $16+17i$

Multiplying and dividing complex numbers

1) $9+40i$
2) $8-20i$
3) $-34-34i$
4) $-60+2i$
5) $25+77i$
6) $25+46i$
7) $\frac{i}{2}$
8) $\frac{3}{4}+i$
9) $9-5i$
10) $\frac{2}{5}-\frac{6}{5}i$
11) $-2+\frac{3}{5}i$
12) $-\frac{9}{10}+\frac{27}{10}i$

Rationalizing imaginary denominators

1) $2+2i$
2) $\frac{3}{2}+\frac{2}{3}i$
3) $\frac{8}{9}+\frac{2}{3}i$
4) $\frac{12}{5}-\frac{4}{5}i$
5) $-\frac{3}{4}-\frac{1}{4}i$
6) $\frac{5}{12}+\frac{5}{4}i$
7) $-\frac{117}{145}-\frac{41}{145}i$
8) $-\frac{5}{149}-\frac{71}{149}i$
9) $-\frac{1}{5}-\frac{1}{5}i$
10) $-1+6i$
11) $-\frac{41}{106}+\frac{11}{106}i$
12) $-\frac{1}{2}-\frac{3}{2}i$

Chapter 6: Matrices

Topics that you'll learn in this chapter:

- ✓ Adding and Subtracting Matrices
- ✓ Matrix Multiplications
- ✓ Finding Determinants of a Matrix

Adding and Subtracting Matrices

Step-by-step guide:

- ✓ A matrix (plural: matrices) is a rectangular array of numbers or variables arranged in rows and columns.
- ✓ We can add or subtract two matrices if they have the same dimensions.
- ✓ For addition or subtraction, add or subtract the corresponding entries, and place the result in the corresponding position in the resultant matrix.

Examples:

1) $[2 \quad -5 \quad -3] + [1 \quad -2 \quad -3] =$

Add the elements in the matching positions:

$[2+1 \quad -5+(-2) \quad -3+(-3)] = [3 \quad -7 \quad -6]$

2) $\begin{bmatrix} 3 & 6 \\ -1 & -3 \\ -5 & -1 \end{bmatrix} + \begin{bmatrix} 0 & -1 \\ 6 & 0 \\ 2 & 3 \end{bmatrix} =$

Add the elements in the matching positions:

$\begin{bmatrix} 3+0 & 6+(-1) \\ (-1)+6 & (-3)+0 \\ (-5)+2 & (-1)+3 \end{bmatrix} = \begin{bmatrix} 3 & 5 \\ 5 & -3 \\ -3 & 2 \end{bmatrix}$

Solve.

1) $\begin{bmatrix} 6 & 8 \\ -14 & 33 \end{bmatrix} - \begin{bmatrix} 12 & 5 \\ -27 & -8 \end{bmatrix} =$

2) $\begin{bmatrix} 16 & -4 \\ -38 & 24 \end{bmatrix} + \begin{bmatrix} 9 & -6 \\ 5 & 2 \end{bmatrix} =$

3) $\begin{bmatrix} 12 & 21 \\ -17 & 33 \end{bmatrix} - \begin{bmatrix} 5 & -8 \\ 2 & 19 \end{bmatrix} =$

4) $\begin{bmatrix} 16 & -4 \\ -38 & 14 \end{bmatrix} + \begin{bmatrix} 9 & -6 \\ 5 & 2 \end{bmatrix} =$

5) $\begin{bmatrix} -5 & 2 & -2 \\ 4 & -2 & 0 \end{bmatrix} - \begin{bmatrix} 6 & -5 & -6 \\ 1 & 3 & -3 \end{bmatrix} =$

6) $\begin{bmatrix} -4n & n+m \\ -2n & -4m \end{bmatrix} + \begin{bmatrix} 4 & -5 \\ 3m & 0 \end{bmatrix} =$

7) $\begin{bmatrix} -6r+t \\ -r \\ 6s \end{bmatrix} + \begin{bmatrix} 6r \\ -4t \\ -3r+2 \end{bmatrix} =$

8) $\begin{bmatrix} z-5 \\ -6 \\ -1-6z \\ 3y \end{bmatrix} + \begin{bmatrix} -3y \\ 3z \\ 5+z \\ 4z \end{bmatrix} =$

Matrix Multiplication

Step-by-step guide:

- ✓ Step 1: Make sure that it's possible to multiply the two matrices (the number of columns in the 1st one should be the same as the number of rows in the second one.)
- ✓ Step 2: The elements of each row of the first matrix should be multiplied by the elements of each column in the second matrix.
- ✓ Step 3: Add the products.

Examples:

1) $\begin{bmatrix} -5 & -5 \\ -1 & 2 \end{bmatrix}\begin{bmatrix} -2 & -3 \\ 3 & 5 \end{bmatrix} =$

Multiply the rows of the first matrix by the columns of the second matrix.

$$\begin{bmatrix} (-5)(-2) + (-5).3 & (-5)(-3) + (-5).5 \\ (-1)(-2) + 2.3 & (-1)(-3) + 2.5 \end{bmatrix} = \begin{bmatrix} -5 & -10 \\ 8 & 13 \end{bmatrix}$$

2) $\begin{bmatrix} -4 & -6 & -6 \\ 0 & 6 & 3 \end{bmatrix}\begin{bmatrix} 0 \\ -3 \\ 0 \end{bmatrix} =$

Multiply the rows of the first matrix by the columns of the second matrix.

$$\begin{bmatrix} (-4).0 + (-6)(-3) + (-6).0 \\ 0.0 + 6(-3) + 3.0 \end{bmatrix} = \begin{bmatrix} 18 \\ -18 \end{bmatrix}$$

Solve.

1) $\begin{bmatrix} 0 & 2 \\ -2 & -5 \end{bmatrix}\begin{bmatrix} 6 & -6 \\ 3 & 0 \end{bmatrix} =$

2) $\begin{bmatrix} 3 & -1 \\ -3 & 6 \\ -6 & -6 \end{bmatrix}\begin{bmatrix} -1 & 6 \\ 5 & 4 \end{bmatrix} =$

3) $\begin{bmatrix} 0 & 5 \\ -3 & 1 \\ -5 & 1 \end{bmatrix}\begin{bmatrix} -4 & 4 \\ -2 & -4 \end{bmatrix} =$

4) $\begin{bmatrix} 5 & 3 & 5 \\ 1 & 5 & 0 \end{bmatrix}\begin{bmatrix} -4 & 2 \\ -3 & 4 \\ 3 & -5 \end{bmatrix} =$

5) $\begin{bmatrix} 4 & 5 \\ -4 & 6 \\ -5 & -6 \end{bmatrix}\begin{bmatrix} 4 & 6 \\ 6 & 2 \\ -4 & 1 \end{bmatrix} =$

6) $\begin{bmatrix} -2 & -6 \\ -4 & 3 \\ 5 & 0 \\ 4 & -6 \end{bmatrix}\begin{bmatrix} 2 & -2 & 2 \\ -2 & 0 & -3 \end{bmatrix} =$

7) $\begin{bmatrix} -1 & 1 & -1 \\ 5 & 2 & -5 \\ 6 & -5 & 1 \\ -5 & 6 & 0 \end{bmatrix}\begin{bmatrix} 6 & 5 \\ 5 & -6 \\ 6 & 0 \end{bmatrix} =$

8) $\begin{bmatrix} 5 & 3 & 2 \\ 6 & 4 & 1 \\ 7 & -9 & 12 \end{bmatrix}\begin{bmatrix} -2 & 5 & 4 \\ 5 & 6 & 13 \\ 3 & 2 & 1 \end{bmatrix} =$

Finding Determinants of a Matrix

Step-by-step guide:

$\begin{bmatrix} a & b \\ c & d \end{bmatrix}$ $\quad |A| = ad - bc$

$\begin{bmatrix} a & b & c \\ d & e & f \\ g & h & i \end{bmatrix}$ $\quad |A| = a(ei - fh) - b(di - fg) + c(dh - eg)$

Examples:

1) Evaluate the determinant of matrix. $\begin{bmatrix} 0 & -4 \\ -6 & -2 \end{bmatrix}$
   Use the matrix determinant: $|A| = ad - bc = (0)(-2) - (-4)(-6) = -24$

2) Evaluate the determinant of matrix. $\begin{bmatrix} 3 & 5 & 1 \\ 1 & 4 & 2 \\ 7 & 1 & 9 \end{bmatrix}$
   Use the matrix determinant: $|A| = a(ei - fh) - b(di - fg) + c(dh - eg)$

   $|A| = 3(4 \times 9 - 2 \times 1) - 5(1 \times 9 - 2 \times 7) + 1(1 \times 1 - 4 \times 7) = 100$

Evaluate the determinant of each matrix.

1) $\begin{bmatrix} 8 & 5 \\ -4 & -6 \end{bmatrix} =$

2) $\begin{bmatrix} 0 & 4 \\ 6 & 5 \end{bmatrix} =$

3) $\begin{bmatrix} 6 & 1 & 7 \\ 2 & -3 & 3 \\ 4 & -1 & 2 \end{bmatrix} =$

4) $\begin{bmatrix} -2 & 5 & -4 \\ 0 & -3 & 5 \\ -5 & 5 & -6 \end{bmatrix} =$

5) $\begin{bmatrix} -3 & 1 & 8 \\ -9 & -1 & 7 \\ 0 & 2 & 1 \end{bmatrix} =$

6) $\begin{bmatrix} 5 & 3 & 3 \\ -4 & -5 & 1 \\ 5 & 3 & 0 \end{bmatrix} =$

7) $\begin{bmatrix} 6 & 2 & -1 \\ -5 & -4 & -5 \\ 3 & -3 & 1 \end{bmatrix} =$

8) $\begin{bmatrix} 6 & 5 & -3 \\ -5 & 4 & -2 \\ 1 & -4 & 5 \end{bmatrix} =$

9) $\begin{bmatrix} -1 & -8 & 9 \\ 4 & 12 & -7 \\ -10 & 3 & 2 \end{bmatrix} =$

10) $\begin{bmatrix} 3 & 9 & 1 \\ 2 & -10 & 1 \\ 5 & 3 & 8 \end{bmatrix} =$

Answers of Worksheets – Chapter 6

Adding and subtracting matrices

1) $\begin{bmatrix} -6 & 3 \\ 13 & 41 \end{bmatrix}$

2) $\begin{bmatrix} 25 & -10 \\ -33 & 16 \end{bmatrix}$

3) $\begin{bmatrix} 7 & 29 \\ -19 & 14 \end{bmatrix}$

4) $\begin{bmatrix} 25 & -10 \\ -33 & 16 \end{bmatrix}$

5) $\begin{bmatrix} -11 & 7 & 4 \\ 3 & -5 & 3 \end{bmatrix}$

6) $\begin{bmatrix} -4n+4 & n+m-5 \\ -2n+3m & -4m \end{bmatrix}$

7) $\begin{bmatrix} t \\ -r-4t \\ 6s-3r+2 \end{bmatrix}$

8) $\begin{bmatrix} z-5-3y \\ -6+3z \\ 4-5z \\ 3y+4z \end{bmatrix}$

Matrix multiplication

1) $\begin{bmatrix} 6 & 0 \\ -27 & 12 \end{bmatrix}$

2) $\begin{bmatrix} -8 & 14 \\ 33 & 6 \\ -24 & -60 \end{bmatrix}$

3) $\begin{bmatrix} -10 & -20 \\ 10 & -16 \\ 18 & -24 \end{bmatrix}$

4) $\begin{bmatrix} -14 & -3 \\ -19 & 22 \end{bmatrix}$

5) Undefined

6) $\begin{bmatrix} 8 & 4 & 14 \\ -14 & 8 & -17 \\ 10 & -10 & 10 \\ 20 & -8 & 26 \end{bmatrix}$

7) $\begin{bmatrix} -7 & -11 \\ 10 & 13 \\ 17 & 60 \\ 0 & -61 \end{bmatrix}$

8) $\begin{bmatrix} 11 & 47 & 61 \\ 11 & 56 & 77 \\ -23 & 5 & -77 \end{bmatrix}$

Finding determinants of a matrix

1) -28

2) -24

3) 60

4) -51

5) 90

6) 39

7) -161

8) 139

9) 647

10) -292

Chapter 7:

Polynomial Operations

Topics that you'll learn in this chapter:

- ✓ Writing Polynomials in Standard Form
- ✓ Simplifying Polynomials
- ✓ Adding and Subtracting Polynomials
- ✓ Multiplying Monomials
- ✓ Multiplying and Dividing Monomials
- ✓ Multiplying a Polynomial and a Monomial
- ✓ Multiplying Binomials

Writing Polynomials in Standard Form

Step-by-step guide:

- ✓ A polynomial function $f(x)$ of degree n is of the form

$$f(x) = a_n x^n + a_{n-1} x_{n-1} + \cdots + a_1 x + a_0$$

- ✓ The first term is the one with the biggest power!

Examples:

1) Write this polynomial in standard form. $-12 + 3x^2 - 6x^4 =$

The first term is the one with the biggest power: $-12 + 3x^2 - 6x^4 = -6x^4 + 3x^2 - 12$

2) Write this polynomial in standard form. $5x^2 - 9x^5 + 8x^3 - 11 =$

The first term is the one with the biggest power: $5x^2 - 9x^5 + 8x^3 - 11 =$
$-9x^5 + 8x^3 + 5x^2 - 11$

Write each polynomial in standard form.

1) $9x - 7x =$
2) $-3 + 16x - 16x =$
3) $3x^2 - 5x^3 =$
4) $3 + 4x^3 - 3 =$
5) $2x^2 + 1x - 6x^3 =$
6) $-x^2 + 2x^3 =$
7) $2x + 4x^3 - 2x^2 =$
8) $-2x^2 + 4x - 6x^3 =$
9) $2x^2 + 2 - 5x =$
10) $12 - 7x + 9x^4 =$
11) $5x^2 + 13x - 2x^3 =$
12) $10 + 6x^2 - x^3$

Simplifying Polynomials

Step-by-step guide:

- ✓ Find "like" terms. (they have same variables with same power).
- ✓ Use "FOIL". (First-Out-In-Last) for binomials:

$$(x+a)(x+b) = x^2 + (b+a)x + ab$$

- ✓ Add or Subtract "like" terms using order of operation.

Examples:

1) Simplify this expression. $4x(6x-3) =$

   Use Distributive Property: $4x(6x-3) = 24x^2 - 12x$

2) Simplify this expression. $(6x-2)(2x-3) =$

   First apply FOIL method: $(a+b)(c+d) = ac + ad + bc + bd$

   $(6x-2)(2x-3) = 12x^2 - 18x - 4x + 6$

   Now combine like terms: $12x^2 - 18x - 4x + 6 = 12x^2 - 22x + 6$

Simplify each expression.

1) $5(2x-10) =$

2) $2x(4x-2) =$

3) $4x(5x-3) =$

4) $3x(7x+3) =$

5) $4x(8x-4) =$

6) $5x(5x+4) =$

7) $(2x-3)(x-4) =$

8) $(x-5)(3x+4) =$

9) $(x-5)(x-3) =$

10) $(3x+8)(3x-8) =$

11) $(3x-8)(3x-4) =$

12) $3x^2 + 3x^2 - 2x^3 =$

Adding and Subtracting Polynomials

Step-by-step guide:

- ✓ Adding polynomials is just a matter of combining like terms, with some order of operations considerations thrown in.
- ✓ Be careful with the minus signs, and don't confuse addition and multiplication!

Examples:

1) *Simplify the expressions.* $(4x^3 + 3x^4) - (x^4 - 5x^3) =$

First use Distributive Property for $-(x^4 - 5x^3)$, $\rightarrow -(x^4 - 5x^3) = -x^4 + 5x^3$

$(4x^3 + 3x^4) - (x^4 - 5x^3) = 4x^3 + 3x^4 - x^4 + 5x^3$

Now combine like terms: $4x^3 + 3x^4 - x^4 + 5x^3 = 2x^4 + 9\quad{}^3$

2) *Add expressions.* $(2x^3 - 6) + (9x^3 - 4x^2) =$

Remove parentheses: $(2x^3 - 6) + (9x^3 - 4x^2) = 2x^3 - 6 + 9x^3 - 4x^2$

Now combine like terms: $2x^3 - 6 + 9x^3 - 4x^2 = 11x^3 - 4x^2 - 6$

✍ *Add or subtract expressions.*

1) $(-x^2 - 2) + (2x^2 + 1) =$

2) $(2x^2 + 3) - (3 - 4x^2) =$

3) $(2x^3 + 3x^2) - (x^3 + 8) =$

4) $(4x^3 - x^2) + (3x^2 - 5x) =$

5) $(7x^3 + 9x) - (3x^3 + 2) =$

6) $(2x^3 - 2) + (2x^3 + 2) =$

7) $(4x^3 + 5) - (7 - 2x^3) =$

8) $(4x^2 + 2x^3) - (2x^3 + 5) =$

9) $(4x^2 - x) + (3x - 5x^2) =$

10) $(7x + 9) - (3x + 9) =$

11) $(4x^4 - 2x) - (6x - 2x^4) =$

12) $(12x - 4x^3) - (8x^3 + 6x) =$

Multiplying Monomials

Step-by-step guide:

✓ A monomial is a polynomial with just one term, like $2x$ or $7y$.

Examples:

1) *Multiply expressions.* $5a^4b^3 \times 2a^3b^2 =$

   Use this formula: $x^a \times x^b = x^{a+b}$

   $a^4 \times a^3 = a^{4+3} = a^7$ and $b^3 \times b^2 = b^{3+2} = b^5$

   Then: $5a^4b^3 \times 2a^3b^2 = 10a^7b^5$

2) *Multiply expressions.* $-4xy^4z^2 \times 3x^2y^5z^3 =$

   Use this formula: $x^a \times x^b = x^{a+b}$

   $x \times x^2 = x^{1+2} = x^3$, $y^4 \times y^5 = y^{4+5} = y^9$ and $z^2 \times z^3 = z^{2+3} = z^5$

   Then: $-4xy^4z^2 \times 3x^2y^5z^3 = -12x^3y^9z^5$

✍ ***Simplify each expression***

1) $4u^9 \times (-2u^3) =$

2) $(-2p^7) \times (-3p^2) =$

3) $3xy^2z^3 \times 2z^2 =$

4) $5u^5t \times 3ut^2 =$

5) $(-9a^6) \times (-5a^2b^4) =$

6) $-2a^3b^2 \times 4a^2b =$

7) $2xy^2 \times x^2y^3 =$

8) $3p^2q^4 \times (-2pq^3) =$

9) $4s^5t^2 \times 4st^3 =$

10) $(-6x^3y^2) \times 3x^2y =$

11) $2xy^2z \times 4z^2 =$

12) $4xy \times x^2y =$

Multiplying and Dividing Monomials

Step-by-step guide:

- ✓ When you divide two monomials you need to divide their coefficients and then divide their variables.
- ✓ In case of exponents with the same base, you need to subtract their powers.
- ✓ Exponent's rules:

$$x^a \times x^b = x^{a+b}, \quad \frac{x^a}{x^b} = x^{a-b}$$
$$\frac{1}{x^b} = x^{-b}, \quad (x^a)^b = x^{a \times b}$$
$$(xy)^a = x^a \times y^a$$

Examples:

1) *Multiply expressions.* $(-3x^7)(4x^3) =$
   Use this formula: $x^a \times x^b = x^{a+b} \rightarrow x^7 \times x^3 = x^{10}$, Then: $(-3x^7)(4x^3) = -12x^{10}$
2) *Dividing expressions.* $\frac{18x^2y^5}{2xy^4} =$
   Use this formula: $\frac{x^a}{x^b} = x^{a-b}$, $\frac{x^2}{x} = x^{2-1} = x$ and $\frac{y^5}{y^4} = y^{5-4} = y$
   Then: $\frac{18x^2y^5}{2xy^4} = 9xy$

Simplify each expression.

1) $(-2x^3y^4)(3x^3y^2) =$
2) $(-5x^3y^2)(-2x^4y^5) =$
3) $(9x^5y)(-3x^3y^3) =$
4) $(8x^7y^2)(6x^5y^4) =$
5) $(7x^4y^6)(4x^3y^4) =$
6) $(12x^2y^9)(7x^9y^{12}) =$
7) $\frac{12x^6y^8}{4x^4y^2} =$
8) $\frac{26x^9y^5}{2x^3y^4} =$
9) $\frac{80x^{12}y^9}{10\ \ ^6y^7} =$
10) $\frac{95x^{18}y^7}{5x^9y^2} =$
11) $\frac{200x^3y^8}{40\ \ ^3y^7} =$
12) $\frac{-15x^{17}y^{13}}{3x^6y^9} =$

Multiplying a Polynomial and a Monomial

Step-by-step guide:

- ✓ When multiplying monomials, use the product rule for exponents.
- ✓ When multiplying a monomial by a polynomial, use the distributive property.

$$a \times (b + c) = a \times b + a \times c$$

Examples:

1) *Multiply expressions.* $-4x(5x + 9) =$

Use Distributive Property: $-4x(5x + 9) = -20x^2 - 36x$

2) *Multiply expressions.* $2x(6x^2 - 3y^2) =$

Use Distributive Property: $2x(6x^2 - 3y^2) = 12x^3 - 6xy^2$

Find each product.

1) $3x(9x + 2y) =$

2) $6x(x + 2y) =$

3) $9x(2x + 4y) =$

4) $12x(3x + 9) =$

5) $11x(2x - 11y) =$

6) $2x(6x - 6y) =$

7) $2x(3x - 6y + 3) =$

8) $5x(3x^2 + 2y^2) =$

9) $13x(4x + 8y) =$

10) $5(2x^2 - 9y^2) =$

11) $3x(-2x^2y + 3y) =$

12) $-2(2x^2 - 2xy + 2) =$

Multiplying Binomials

Step-by-step guide:

✓ Use "FOIL". (First-Out-In-Last)

$$(x+a)(x+b) = x^2 + (b+a)x + ab$$

Examples:

1) *Multiply Binomials.* $(x-2)(x+2) =$

Use "FOIL". (First–Out–In–Last): $(x-2)(x+2) = x^2 + 2x - 2x - 4$

Then simplify: $x^2 + 2x - 2x - 4 = x^2 - 4$

2) *Multiply Binomials.* $(x+5)(x-2) =$

Use "FOIL". (First–Out–In–Last):

$(x+5)(x-2) = x^2 - 2x + 5x - 10$

Then simplify: $x^2 - 2x + 5x - 10 = x^2 + 3x - 10$

Find each product.

1) $(x+2)(x+2) =$
2) $(x-3)(x+2) =$
3) $(x-2)(x-4) =$
4) $(x+3)(x+2) =$
5) $(x-4)(x-5) =$
6) $(x+5)(x+2) =$
7) $(x-6)(x+3) =$
8) $(x-8)(x-4) =$
9) $(x+2)(x+8) =$
10) $(x-2)(x+4) =$
11) $(x+4)(x+4) =$
12) $(x+5)(x+5) =$

Factoring Trinomials

Step-by-step guide:

- ✓ "FOIL":

$$(x+a)(x+b) = x^2 + (b+a)x + ab$$

- ✓ "Difference of Squares":

$$a^2 - b^2 = (a+b)(a-b)$$

$$a^2 + 2ab + b^2 = (a+b)(a+b)$$

$$a^2 - 2ab + b^2 = (a-b)(a-b)$$

- ✓ "Reverse FOIL":

$$x^2 + (b+a)x + ab = (x+a)(x+b)$$

Examples:

1) *Factor this trinomial.* $x^2 - 2x - 8 =$

   Break the expression into groups: $(x^2 + 2x) + (-4x - 8)$

   Now factor out x from $x^2 + 2x$: $x(x+2)$ and factor out -4 from $-4x - 8$: $-4(x+2)$

   Then: $= x(x+2) - 4(x+2)$, now factor out like term: $x + 2$

   Then: $(x+2)(x-4)$

2) *Factor this trinomial.* $x^2 - 6x + 8 =$

   Break the expression into groups: $(x^2 - 2x) + (-4x + 8)$

   Now factor out x from $x^2 - 2x$: $x(x-2)$, and factor out -4 from $-4x + 8$: $-4(x-2)$

   Then: $= x(x-2) - 4(x-2)$, now factor out like term: $x - 2$

   Then: $(x-2)(x-4)$

Factor each trinomial.

1) $x^2 + 8x + 15 =$

2) $x^2 - 5x + 6 =$

3) $x^2 + 6x + 8 =$

4) $x^2 - 8x + 16 =$

5) $x^2 - 7x + 12 =$

6) $x^2 + 11x + 18 =$

7) $x^2 + 2x - 24 =$

8) $x^2 + 4x - 12 =$

9) $x^2 - 10x + 9 =$

10) $x^2 + 5x - 14 =$

11) $x^2 - 6x - 27 =$

12) $x^2 - 11x - 42 =$

Answers of Worksheets – Chapter 7

Writing Polynomials in Standard Form

1) $2x$
2) -3
3) $-5x^3 + 3x^{-2}$
4) $4x^3$
5) $-6x^3 + 2x^2 + x$
6) $2x^3 - x^2$
7) $4x^3 - 2x^2 + 2x$
8) $-6x^3 - 2x^2 + 4x$
9) $2x^2 - 5x + 2$
10) $9x^4 - 7x + 12$
11) $-2x^3 + 5x^2 + 13x$
12) $-x^3 + 6x^2 + 10$

Simplifying Polynomials

1) $10x - 50$
2) $8x^2 - 4x$
3) $20x^2 - 12x$
4) $21x^2 + 9x$
5) $32x^2 - 16x$
6) $25x^2 + 20x$
7) $2x^2 - 11x + 12$
8) $3x^2 - 11x - 20$
9) $x^2 - 8x + 15$
10) $9x^2 - 64$
11) $9x^2 - 36x + 32$
12) $-2x^3 + 6x^2$

Adding and Subtracting Polynomials

1) $x^2 - 1$
2) $6x^2$
3) $x^3 + 3x^2 - 8$
4) $4x^3 + 2x^2 - 5x$
5) $4x^3 + 9x - 2$
6) $4x^3$
7) $6x^3 - 2$
8) $4x^2 - 5$
9) $-x^2 + 2x$
10) $4x$
11) $6x^4 - 8x$
12) $-12x^3 + 6x$

Multiplying Monomials

1) $-8u^{12}$
2) $6p^9$
3) $6xy^2z^5$
4) $15u^6t^3$
5) $45a^8b^4$
6) $-8a^5b^3$
7) $2x^3y^5$
8) $-6p^3q^7$
9) $16s^6t^5$
10) $-18x^5y^3$
11) $8xy^2z^3$
12) $4x^3y^2$

Multiplying and Dividing Monomials

1) $-6x^6y^6$
2) $10x^7y^7$
3) $-27x^8y^4$
4) $48x^{12}y^6$
5) $28x^7y^{10}$
6) $84x^{11}y^{21}$
7) $3x^2y^6$
8) $13x^6y$
9) $8x^6y^2$
10) $19x^9y^5$
11) $5y$
12) $-5x^{11}y^4$

Multiplying a Polynomial and a Monomial

1) $27x^2 + 6xy$
2) $6x^2 + 12xy$
3) $18x^2 + 36xy$
4) $36x^2 + 108x$
5) $22x^2 - 121xy$
6) $12x^2 - 12xy$
7) $6x^2 - 12xy + 6x$
8) $15x^3 + 10xy^2$
9) $52x^2 + 104xy$
10) $10x^2 - 45y^2$
11) $-6x^3y + 9xy$
12) $-4x^2 + 4xy - 4$

Multiplying Binomials

1) $x^2 + 4x + 4$
2) $x^2 - x - 6$
3) $x^2 - 6x + 8$
4) $x^2 + 5x + 6$
5) $x^2 - 9x + 20$
6) $x^2 + 7x + 10$
7) $x^2 - 3x - 18$
8) $x^2 - 12x + 32$
9) $x^2 + 10x + 16$
10) $x^2 + 2x - 8$
11) $x^2 + 8x + 16$
12) $x^2 + 10x + 25$

Factoring Trinomials

1) $(x+3)(x+5)$
2) $(x-2)(x-3)$
3) $(x+4)(x+2)$
4) $(x-4)(x-4)$
5) $(x-3)(x-4)$
6) $(x+2)(x+9)$
7) $(x+6)(x-4)$
8) $(x-2)(x+6)$
9) $(x-1)(x-9)$
10) $(x-2)(x+7)$
11) $(x-9)(x+3)$
12) $(x+3)(x-14)$

Chapter 8:

Functions Operations

Topics that you'll learn in this chapter:

- ✓ Function notation
- ✓ Adding and subtracting functions
- ✓ Multiplying and dividing functions
- ✓ Composition of functions

Function Notation

Step-by-step guide:

- ✓ Functions are mathematical operations that assign unique outputs to given inputs.
- ✓ Function notation is the way a function is written. It is meant to be a precise way of giving information about the function without a rather lengthy written explanation.
- ✓ The most popular function notation is $f(x)$ which is read "f of x".

Examples:

1) Evaluate: $w(x) = 3x + 1$, find $w(4)$. Substitute x with 4: Then: $w(x) = 3x + 1 \rightarrow w(4) = 3(4) + 1 \rightarrow w(x) = 12 + 1 \rightarrow w(x) = 13$
2) Evaluate: $h(n) = n^2 - 10$, find $h(-2)$. Substitute x with -2:

   Then: $h(n) = n^2 - 10 \rightarrow h(-2) = (-2)^2 - 10 \rightarrow h(-2) = 4 - 10 \rightarrow h(-2) = -6$

Evaluate each function.

1) $f(x) = -x + 5$, find $f(4)$
2) $g(n) = 10n - 3$, find $g(6)$
3) $g(n) = 8n + 4$, find $g(1)$
4) $h(x) = 4x - 22$, find $h(2)$
5) $h(a) = -11a + 5$, find $h(2a)$
6) $k(a) = 7a + 3$, find $k(a - 2)$
7) $h(x) = 3x + 5$, find $h(6x)$
8) $h(n) = n^2 - 10$, find $h(5)$
9) $h(n) = -2n^2 - 6n$, find $h(2)$
10) $g(n) = 3n^2 + 2n$, find $g(2)$
11) $h(x) = x^2 + 1$, find $h(\frac{x}{4})$
12) $h(x) = x^3 + 8$, find $h(-2)$

Adding and Subtracting Functions

Step-by-step guide:

- ✓ Just like we can add and subtract numbers, we can add and subtract functions. For example, if we had functions f and g, we could create two new functions:
- ✓ f + g and f - g.

Examples:

1) $f(x) = 2x + 4$, $g(x) = x + 3$, Find: $(f - g)(1)$

   $(f - g)(x) = f(x) - g(x)$, then: $(f - g)(x) = 2x + 4 - (x + 3)$

   $= 2x + 4 - x - 3 = x + 1$

   Substitute x with 1: $(f - g)(1) = 1 + 1 = 2$

2) $g(a) = 2a - 1$, $f(a) = -a - 4$, Find: $(g + f)(-1)$

   $(g + f)(a) = g(a) + f(a)$, Then: $(g + f)(a) = 2a - 1 - a - 4 = a - 5$

   Substitute a with -1: $(g + f)(a) = a - 5 = -1 - 5 = -6$

Perform the indicated operation.

1) $g(x) = 2x - 5$

   $h(x) = 4x + 5$

   Find: $g(3) - h(3)$

2) $h(3) = 3x + 3$

   $g(x) = -4x + 1$

   Find: $(h + g)(10)$

3) $f(x) = 4x - 3$

   $g(x) = x^3 + 2x$

   Find: $(f - g)(4)$

4) $h(n) = 4n + 5$

   $g(n) = 3n + 4$

   Find: $(h - g)(n)$

5) $g(x) = -x^2 - 1 - 2x$

   $f(x) = 5 + x$

   Find: $(g - f)(x)$

6) $g(t) = 2t + 5$

   $f(t) = -t^2 + 5$

   Find: $(g + f)(t)$

Multiplying and Dividing Functions

Step-by-step guide:

- ✓ Just like we can multiply and divide numbers, we can multiply and divide functions. For example, if we had functions f and g, we could create two new functions: f × g, and $\frac{f}{g}$.

Examples:

1) $g(x) = -x - 2, f(x) = 2x + 1$, Find: $(g.f)(2)$

$(g.f)(x) = g(x).f(x) = (-x-2)(2x+1) = -2x^2 - x - 4x - 2 = -2x^2 - 5x - 2$

Substitute x with 2:

$(g.f)(x) = -2x^2 - 5x - 2 = -2(2)^2 - 5(2) - 2 = -8 - 10 - 2 = -20$

2) $f(x) = x + 4, h(x) = 5x - 2$, Find: $\left(\frac{f}{h}\right)(-1)$

$\left(\frac{f}{h}\right)(x) = \frac{f(x)}{h(x)} = \frac{x+4}{5x-2}$

Substitute x with -1: $\left(\frac{f}{h}\right)(x) = \frac{x+4}{5x-2} = \frac{(-1)+4}{5(-1)-2} = \frac{3}{-7} = -\frac{3}{7}$

Perform the indicated operation.

1) $f(x) = 2a^2$
   $g(x) = -5 + 3a$
   Find $(\frac{f}{g})(2)$

2) $g(a) = 3a + 2$
   $f(a) = 2a - 4$
   Find $(\frac{g}{f})(3)$

3) $g(t) = t^2 + 3$
   $h(t) = 4t - 3$
   Find $(g.h)(-1)$

4) $g(n) = n^2 + 4 + 2n$
   $h(n) = -3n + 2$
   Find $(g.h)(1)$

5) $f(x) = 2x^3 - 5x^2$
   $g(x) = 2x - 1$
   Find $(f.g)(x)$

6) $f(x) = 3x - 1$
   $g(x) = x^2 - x$
   Find $(\frac{f}{g})(x)$

Composition of Functions

Step-by-step guide:

- ✓ The term "composition of functions" (or "composite function") refers to the combining together of two or more functions in a manner where the output from one function becomes the input for the next function.
- ✓ The notation used for composition is: $(f\ o\ g)(x) = f(g(x))$

Examples:

1) *Using* $\mathrm{f(x) = x + 2}$ *and* $\mathrm{g(x) = 4x}$, *find:* $\boldsymbol{f(g(1))}$

$(f\ o\ g)(x) = f(g(x))$

Then: $(f\ o\ g)(x) = f\big(g(x)\big) = f(4x) = 4x + 2$

Substitute x with 1: $(f\ o\ g)(1) = 4 + 2 = 6$

2) *Using* $\mathrm{f(x) = 5x + 4}$ *and* $\mathrm{g(x) = x - 3}$, *find:* $\boldsymbol{g(f(3))}$

$(f\ o\ g)(x) = f(g(x))$

Then: $(g\ o\ f)(x) = g\big(f(x)\big) = g(\mathrm{5x + 4})$, *now substitute* x *in* $\mathrm{g(x)}$ *by* $\mathrm{5x + 4}$. *Then:* $g(\mathrm{5x} + 4) = (5x + 4) - 3 = 5x + 4 - 3 = 5x + 1$

Substitute x with 3: $(g\ o\ f)(x) = g\big(f(x)\big) = 5x + 1 = 5(3) + 1 = 15 = 1 = 16$

Using $\mathbf{f(x) = 6x + 2}$ ***and*** $\mathbf{g(x) = x - 5}$, ***find:***

1) $f(g(7))$

2) $f(f(2))$

3) $g(f(3))$

4) $g(g(x))$

Using $\mathbf{f(x) = 7x + 4}$ ***and*** $\mathbf{g(x) = 2x - 4}$, ***find:***

5) $f(g(3))$

6) $f(f(3))$

7) $g(f(4))$

8) $g(g(5))$

Answers of Worksheets – Chapter 8

Function Notation

1) 1
2) 57
3) 12
4) -14
5) $-22a+5$
6) $7a-11$
7) $18x+5$
8) 15
9) -20
10) 16
11) $1+\frac{1}{16}x^2$
12) 0

Adding and Subtracting Functions

1) -16
2) -6
3) -59
4) $n+1$
5) $-x^2-3x-6$
6) $-t^2+2t+10$

Multiplying and Dividing Functions

1) 8
2) $\frac{11}{2}$
3) -28
4) -7
5) $4x^4-12x^3+5x^2$
6) $\frac{3x-1}{x^2-x}$

Composition of functions

1) 14
2) 86
3) 15
4) $x-10$
5) 18
6) 179
7) 60
8) 8

Chapter 9: Logarithms

Topics that you'll learn in this chapter:

- ✓ Evaluating Logarithms
- ✓ Properties of Logarithms
- ✓ Natural Logarithms
- ✓ Solving Logarithmic Equations

Evaluating Logarithms

Step-by-step guide:

✓ Logarithm is another way of writing exponent. $log_b\, y = x$ is equivalent to $y = b^x$

✓ Learn some logarithms rules:

$$\boldsymbol{log_b(x) = \frac{log_d(x)}{log_d(b)}} \qquad \boldsymbol{log_a x^b = b\,log_a x}$$

$$\boldsymbol{log_a a = 1} \qquad \boldsymbol{log_a 1 = 0}$$

Examples:

1) *Evaluate:* $log_4 64$

Rewrite 64 in power base form: $64 = 4^3$, then: $log_4 64 = log_4(4^3)$

Use log rule: $log_a(x^b) = b.log_a(x) \rightarrow log_4(4^3) = 3log_4(4)$

Use log rule: $log_a(a) = 1 \rightarrow log_4(4) = 1.$ $\quad 3log_4(4) = 3 \times 1 = 3$

2) *Evaluate:* $log_5 625$

Rewrite 625 in power base form: $625 = 5^4$, then: $log_5 625 = log_5(5^4)$

Use log rule: $log_a(x^b) = b.log_a(x) \rightarrow log_5(5^4) = 4log_5(5)$

Use log rule: $log_a(a) = 1 \rightarrow log_5(5) = 1.$ $\quad 4log_5(5) = 4 \times 1 = 4$

Evaluate each logarithm.

1) $log_2 \frac{1}{2} =$

2) $log_2 \frac{1}{8} =$

3) $log_3 \frac{1}{3} =$

4) $log_4 \frac{1}{16} =$

Circle the points which are on the graph of the given logarithmic functions.

5) $y = 2log_3(x+1) + 2$	$(2,4)$,	$(8,4)$,	$(0,3)$
6) $y = 3log_3(3x) - 2$	$(3,6)$,	$(3,4)$,	$(\frac{1}{3},2)$
7) $y = -2log_2 2(x-1) + 1$	$(3,-3)$,	$(2,1)$,	$(5,5)$
8) $y = 4log_4(4x) + 7$	$(1,7)$,	$(1,11)$,	$(4,8)$

Properties of Logarithms

Step-by-step guide:

✓ Learn some logarithms properties:

$a^{\log_a b} = b$

$\log_a 1 = 0$

$\log_a a = 1$

$\log_a(x \, . \, y) = \log_a x + \log_a y$

$\log_a \frac{x}{y} = \log_a x - \log_a y$

$\log_a \frac{1}{x} = -\log_a x$

$\log_a x^p = p \, \log_a x$

$\log_{x^k} x = \frac{1}{x} \log_a x, \text{for } k \neq 0$

$\log_a x = \log_{a^c} x^c$

$\log_a x = \frac{1}{\log_x a}$

Examples:

1) *Expand this logarithm.* $\log(8 \times 5) =$

Use log rule: $\log_a(x \, . \, y) = \log_a x + \log_a y$

Then: $\log(8 \times 5) = \log 8 + \log 5$

2) Condense this expression to a single logarithm. $\log 2 - \log 9 =$

Use log rule: $\log_a x - \log_a y = \log_a \frac{x}{y}$

Then: $\log 2 - \log 9 = \log \frac{2}{9}$

✍ ***Expand each logarithm.***

1) $\log(\frac{3}{4}) =$

2) $\log(\frac{5}{7}) =$

3) $\log(\frac{2}{5})^3 =$

4) $\log(2 \times 3^4) =$

5) $\log(\frac{5}{7})^4 =$

6) $\log\left(\frac{2^3}{7}\right) =$

✍ ***Condense each expression to a single logarithm.***

7) $5 \log 6 - 3 \log 4 =$

8) $4 \log 7 - 2 \log 9 =$

9) $3 \log 5 - \log 14 =$

10) $7 \log 3 - 4 \log 4 =$

11) $\log 7 - 2 \log 12 =$

12) $2 \log 5 + 3 \log 8 =$

Natural Logarithms

Step-by-step guide:

- ✓ A natural logarithm is a logarithm that has a special base of the mathematical constant e, which is an irrational number approximately equal to 2.71.
- ✓ The natural logarithm of x is generally written as $\ln x$, or $\log_e x$.

Examples:

1) *Solve the equation for* x: $e^x = 3$

If $f(x) = g(x), then: ln(f(x)) = ln(g(x)) \rightarrow ln(e^x) = ln(3)$

Use log rule: $log_a x^b = b\, log_a x \rightarrow ln(e^x) = x\, ln(e) \rightarrow x ln(e) = ln(3)$

$ln(e) = 1$, then: $x = ln(3)$

2) *Solve equation for* x: $ln(2x - 1) = 1$

Use log rule: $a = \log_b(b^a) \rightarrow 1 = \ln(e^1) = \ln(e) \rightarrow \ln(2x - 1) = \ln(e)$

When the logs have the same base: $\log_b(f(x)) = \log_b(g(x)) \rightarrow f(x) = g(x)$

$ln(2x - 1) = ln(e)$, then: $2x - 1 = e \rightarrow x = \frac{e+1}{2}$

***Solve each equation for* x.**

1) $e^x = 4$

2) $e^x = 8$

3) $ln\, x = 6$

4) $ln\,(ln\, x) = 5$

5) $e^x = 9$

6) $ln(2x + 5) = 4$

7) $ln(6x - 1) = 1$

8) $ln\, x = \frac{1}{2}$

Reduce the following expressions to simplest form.

9) $e^{-2ln5+2ln3} =$

10) $e^{-ln(\frac{1}{e})} =$

11) $2\, ln(e^3) =$

12) $ln(\frac{1}{e})^2 =$

Solving Logarithmic Equations

Step-by-step guide:

- ✓ Convert the logarithmic equation to an exponential equation when it's possible. (If no base is indicated, the base of the logarithm is 10)
- ✓ Condense logarithms if you have more than one log on one side of the equation.
- ✓ Plug in the answers back into the original equation and check to see the solution works.

Examples:

1) *Find the value of the variables in each equation.* $log_4(20 - x^2) = 2$

   Use log rule: $log_b x = log_b y \; then: x = y$

   $2 = log_4(4^2)\,,\, log_4(20 - x^2) = log_4(4^2) = log_4 16$

   Then: $20 - x^2 = 16 \rightarrow 20 - 16 = x^2 \rightarrow x^2 = 4 \rightarrow x = 2$

2) *Find the value of the variables in each equation.* $log(2x + 2) = log(4x - 6)$

   When the logs have the same base: $f(x) = g(x), then: ln(f(x)) = ln(g(x))$

   $log(2x + 2) = log(4x - 6) \rightarrow 2x + 2 = 4x - 6 \rightarrow 2x + 2 - 4x + 6 = 0$

   $2x + 2 - 4x + 6 = 0 \rightarrow -2x + 8 = 0 \rightarrow -2x = -8 \rightarrow x = \frac{-8}{-2} = 4$

✍ *Find the value of the variables in each equation.*

1) $2log_7 - 2x = 0$
2) $-log_5 7x = 2$
3) $\log x + 5 = 2$
4) $log\, x - log\, 4 = 3$
5) $log\, x + log\, 2 = 4$
6) $log\, 10 + log\, x = 1$
7) $log\, x + log\, 8 = log\, 48$
8) $-3\, log_3(x - 2) = -12$
9) $log\, 6x = log\, (x + 5)$
10) $log\, (4k - 5) = log\, (2k - 1)$
11) $log(4p - 2) = log(-5p + 5)$
12) $-10 + log_3\, (n + 3) = -10$

Answers of Worksheets – Chapter 9

Evaluating logarithms

1) -1
2) -3
3) -1
4) -2
5) $(2,4)$
6) $(3,4)$
7) $(3,-3)$
8) $(1,11)$

Properties of logarithms

1) $log\, 3 - log\, 4$
2) $log\, 5 - log\, 7$
3) $3\, log\, 2 - 3\, log\, 5$
4) $log\, 2 + 4\, log\, 3$
5) $4 \log 5 - 4 \log 7$
6) $3log2 - log\, 7$
7) $log\, \frac{6^5}{4^3}$
8) $log\, \frac{7^4}{9^2}$
9) $log\, \frac{5^3}{14}$
10) $log\, \frac{3^7}{4^4}$
11) $log\, \frac{7}{12^2}$
12) $log\, (5^2 8^3)$

Natural logarithms

1) $x = ln\, 4, x = 2ln(2)$
2) $x = ln\, 8, x = 3ln(2)$
3) $x = e^6$
4) $x = e^{e^5}$
5) $x = ln\, 9, x = 2ln(3)$
6) $x = \frac{e^4 - 5}{2}$
7) $x = \frac{e + 1}{6}$
8) $x = \sqrt{e}$
9) $\frac{9}{25} = 0.36$
10) e
11) 6
12) -2

Solving logarithmic equations

1) $\{-\frac{1}{2}\}$
2) $\{\frac{1}{175}\}$
3) $\{\frac{1}{1{,}000}\}$
4) $\{4{,}000\}$
5) $\{5{,}000\}$
6) $\{1\}$
7) $\{6\}$
8) $\{83\}$
9) $\{1\}$
10) $\{2\}$
11) $\{\frac{7}{9}\}$
12) $\{-2\}$

Chapter 10: Radical Expressions

Topics that you'll learn in this chapter:

- ✓ Simplifying Radical Expressions
- ✓ Simplifying Radical Expressions Involving Fractions
- ✓ Multiplying Radical Expressions
- ✓ Adding and Subtracting Radical Expressions
- ✓ Domain and Range of Radical Functions
- ✓ Radical Equations

Simplifying Radical Expressions

Step-by-step guide:

For square roots:

- ✓ Find the prime factors of the numbers inside the radical.
- ✓ Find the largest perfect score factor of the number.
- ✓ Rewrite the radical as the product of perfect score and its matching factor and simplify.

Examples:

1) Find the square root of $\sqrt{225}$.
   First factor the number: $225 = 15^2$, Then: $\sqrt{225} = \sqrt{15^2}$

   Now use radical rule: $\sqrt[n]{a^n} = a$, Then: $\sqrt{15^2} = 15$

2) Evaluate. $\sqrt{4} \times \sqrt{16} =$
   First factor the numbers: $4 = 2^2$ and $16 = 4^2$

   Then: $\sqrt{4} \times \sqrt{16} = \sqrt{2^2} \times \sqrt{4^2}$

   Now use radical rule: $\sqrt[n]{a^n} = a$, Then: $\sqrt{2^2} \times \sqrt{4^2} = 2 \times 4 = 8$

Simplify.

1) $\sqrt{512p^3} =$
2) $\sqrt{216m^4} =$
3) $\sqrt{264x^3y^3} =$
4) $\sqrt{49x^3y^3} =$
5) $\sqrt{16a^4b^3} =$
6) $\sqrt{20x^3y^3} =$
6) $\sqrt[3]{216yx^3} =$
7) $3\sqrt{75x^2} =$
8) $5\sqrt{80x^2} =$
9) $\sqrt[3]{256x^2y^3} =$
10) $\sqrt[3]{343x^4y^2} =$
11) $4\sqrt{125a} =$

Simplifying Radical Expressions Involving Fractions

Step-by-step guide:

- ✓ Radical expressions cannot be in the denominator. (number in the bottom)
- ✓ To get rid of the radical in the denominator, multiply both numerator and denominator by the radical in the denominator.
- ✓ If there is a radical and another integer in the denominator, multiply both numerator and denominator by the conjugate of the denominator.
- ✓ The conjugate of a + b is a-b and vice versa.

Examples:

1) *Simplify* $\frac{3}{\sqrt{5}-3}$

Multiply by the conjugate: $\frac{\sqrt{5}+3}{\sqrt{5}+3} \rightarrow \frac{3}{\sqrt{5}-3} \times \frac{\sqrt{5}+3}{\sqrt{5}+3}$

$(\sqrt{5}-3)(\sqrt{5}+3) = -4$ then: $\frac{3(\sqrt{5}+3)}{-4}$

Use the fraction rule: $\frac{a}{-b} = -\frac{a}{b} \rightarrow \frac{3(\sqrt{5}+3)}{-4} = -\frac{3(\sqrt{5}+3)}{4}$

2) *Simplify* $\frac{9}{\sqrt{7}-3}$

Multiply by the conjugate: $\frac{\sqrt{7}+3}{\sqrt{7}+3}$

$\frac{9}{\sqrt{7}-3} \times \frac{\sqrt{7}+3}{\sqrt{7}+3} = \frac{9\sqrt{7}+3}{-2}$, Use the fraction rule: $\frac{a}{-b} = -\frac{a}{b} \rightarrow \frac{9\sqrt{7}+3}{-2} = -\frac{9\sqrt{7}+3}{2}$

✍ *Simplify.*

1) $\frac{2\sqrt{5r}}{\sqrt{m^3}}$

2) $\frac{8\sqrt{3}}{\sqrt{k}}$

3) $\frac{\sqrt{a}}{\sqrt{a}+\sqrt{b}}$

4) $\frac{1+\sqrt{2}}{3+\sqrt{5}}$

5) $\frac{2+\sqrt{5}}{6-\sqrt{3}}$

6) $\frac{2}{3+\sqrt{5}}$

7) $\frac{\sqrt{7}-\sqrt{3}}{\sqrt{3}-\sqrt{7}}$

8) $\frac{\sqrt{7}+\sqrt{5}}{\sqrt{5}+\sqrt{2}}$

9) $\frac{3\sqrt{2}-\sqrt{7}}{4\sqrt{2}+\sqrt{5}}$

10) $\frac{5\sqrt{3}-3\sqrt{2}}{3\sqrt{2}-2\sqrt{3}}$

11) $\frac{\sqrt{31\ ^5b^3}}{\sqrt{2ab^2}}$

12) $\frac{6\sqrt{45\ ^3}}{3\sqrt{5x}}$

Multiplying Radical Expressions

Step-by-step guide:

- ✓ To multiply radical expressions:
- ✓ Multiply the numbers outside of the radicals.
- ✓ Multiply the numbers inside the radicals.
- ✓ Simplify if needed.

Examples:

1) *Evaluate.* $\sqrt{4} \times \sqrt{16} =$
   First factor the numbers: $4 = 2^2$ and $16 = 4^2$
   Then: $\sqrt{4} \times \sqrt{16} = \sqrt{2^2} \times \sqrt{4^2}$
   Now use radical rule: $\sqrt[n]{a^n} = a$, Then: $\sqrt{2^2} \times \sqrt{4^2} = 2 \times 4 = 8$
2) *Evaluate.* $2\sqrt{3} \times 4\sqrt{2} =$
   Multiply the numbers: $2 \times 4 = 8$
   $2\sqrt{3} \times 4\sqrt{2} = 8\sqrt{3}\sqrt{2}$
   Use radical rule: $\sqrt{a}\sqrt{b} = \sqrt{ab} \rightarrow 8\sqrt{3}\sqrt{2} = 8\sqrt{3 \times 2} = 8\sqrt{6}$
   $= 2 \times 4 \times \sqrt{3 \times 2} = 8\sqrt{6}$

Simplify.

1) $5\sqrt{45} \times 3\sqrt{176} =$
2) $\sqrt{12}(3 + \sqrt{3}) =$
3) $\sqrt{23x^2} \times \sqrt{23x} =$
4) $-5\sqrt{12} \times -\sqrt{3} =$
5) $2\sqrt{20x^2} \times \sqrt{5x^2} =$
6) $\sqrt{12x^2} \times \sqrt{2x^3} =$
7) $-12\sqrt{7x} \times \sqrt{5x^3} =$
8) $-5\sqrt{9x^3} \times 6\sqrt{3x^2} =$
9) $-2\sqrt{12}(3 + \sqrt{12}) =$
10) $\sqrt{18x}\,(4 - \sqrt{6x}) =$
11) $\sqrt{3x}(6\sqrt{x^3} + \sqrt{27}) =$
12) $\sqrt{15r}\,(5 + \sqrt{5}) =$

Adding and Subtracting Radical Expressions

Step-by-step guide:

- ✓ Only numbers that have the same radical part can be added or subtracted.
- ✓ Remember, combining "unlike" radical terms is not possible.
- ✓ For number with the same radical part, just add or subtract factors outside the radicals.

Examples:

1) *Simplify* $4\sqrt{2} + 6\sqrt{2} =$

Add like terms: $4\sqrt{2} + 6\sqrt{2} = 10\sqrt{2}$

2) *Simplify* $\sqrt{3} + 4\sqrt{3} =$

Add like termsâ$\sqrt{3} + 4\sqrt{3} = 1\sqrt{3} + 4\sqrt{3}$

$= (1 + 4)\sqrt{3} = 5\sqrt{3}$

✍ ***Simplify.***

1) $5\sqrt{10} + 4\sqrt{10}$

2) $4\sqrt{12} - 3\sqrt{27}$

3) $-3\sqrt{22} - 4\sqrt{22}$

4) $-10\sqrt{7} + 12\sqrt{7}$

5) $5\sqrt{3} - \sqrt{27}$

6) $-\sqrt{12} + 3\sqrt{3}$

7) $-3\sqrt{6} + 3\sqrt{6}$

8) $3\sqrt{8} + 3\sqrt{2}$

9) $2\sqrt{45} - 2\sqrt{5}$

10) $7\sqrt{18} - 3\sqrt{2}$

11) $-12\sqrt{35} + 14\sqrt{35}$

12) $-6\sqrt{19} - 6\sqrt{19}$

Domain and Range of Radical Functions

Step-by-step guide:

- ✓ To find domain and rage of radical functions, remember that having a negative number under the square root symbol is not possible. (for square roots)
- ✓ To find the domain of the function, find all possible values of the variable inside radical.
- ✓ To find the range, plugin the minimum and maximum values of the variable inside radical.

Examples:

Find the domain and range of the radical function. $y = \sqrt{x-5}$

For domain: Find non-negative values for radicals: $x \geq 5$

$\sqrt{f(x)} \rightarrow f(x) \geq 0$

Then solve $x - 5 \geq 0 \rightarrow x \geq 5$

domain: $x \geq 5$

for range: the range of an radical function of the form $c\sqrt{ax+b} + k$ is $f(x) \geq k$

$k = 0$ then: $f(x) \geq 0$

Identify the domain and range of each

1) $y = \sqrt{x+2} - 3$

2) $y = \sqrt{x-1} - 1$

3) $y = \sqrt{x-2} + 5$

4) $y = \sqrt{x+1} - 4$

Sketch the graph of each function

5) $y = \sqrt{x} + 8$

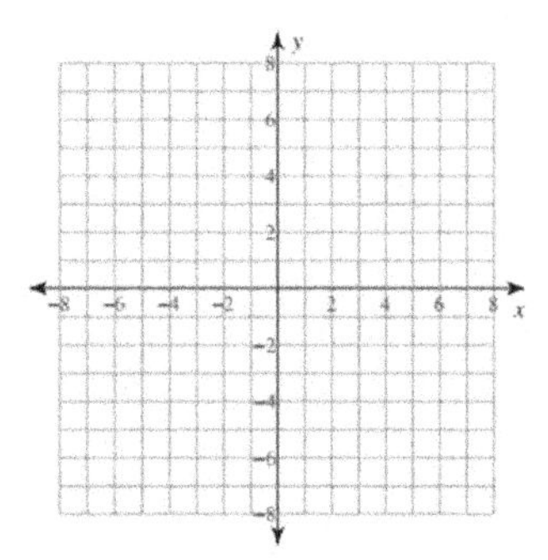

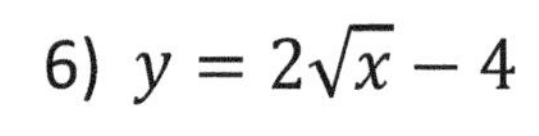

6) $y = 2\sqrt{x} - 4$

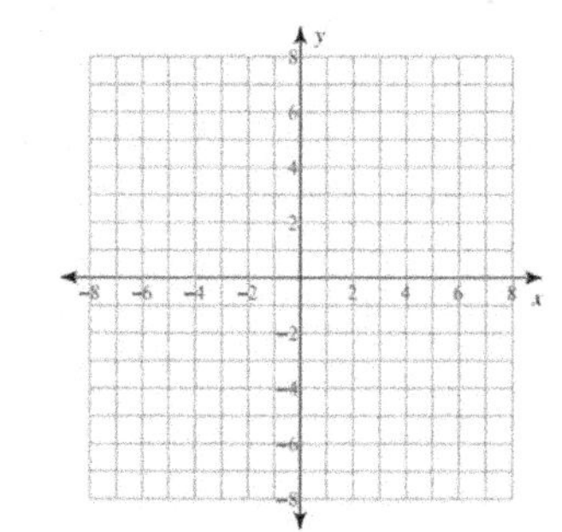

Radical Equations

Step-by-step guide:

- ✓ Isolate the radical on one side of the equation.
- ✓ Square both sides of the equation to remove the radical
- ✓ Solve the equation for the variable
- ✓ Plugin the answer into the original equation to avoid extraneous values.

Examples:

1) Solve $2\sqrt{x+2} = 10$

   Divide both sides by 2: $\frac{2\sqrt{(x+2)}}{2} = \frac{10}{2} \rightarrow \sqrt{x+2} = 5$

   Square both sides: $(\sqrt{(x+2)})^2 = 5^2 \rightarrow x+2 = 25 \rightarrow x = 23$

2) Solve $\sqrt{x} - 3 = 5$

   First square both sides: $(\sqrt{x} - 3)^2 = 5^2$

   Solve: $x - 3 = 25 \rightarrow x = 28$

Solve each equation. Remember to check for extraneous solutions.

1) $\sqrt{(x-6)} = 8$
2) $5 = \sqrt{(x-3)}$
3) $\sqrt{r} = 4$
4) $\sqrt{(m+8)} = 8$
5) $5\sqrt{3x} = 15$
6) $1 = \sqrt{x-5}$
7) $-12 = -6\sqrt{r+4}$
8) $20 = 2\sqrt{36v}$
9) $\sqrt{n+3} - 1 = 7$
10) $\sqrt{3r} = \sqrt{(3r-1)}$
11) $\sqrt{(3x+12)} = \sqrt{(x+8)}$
12) $\sqrt{v} = \sqrt{(2v-6)}$

Answers of Worksheets – Chapter 10

Simplifying radical expressions

1) $16p\sqrt{2p}$
2) $6m^2\sqrt{6}$
3) $2x.y\sqrt{66xy}$
4) $7xy\sqrt{xy}$
5) $4a^2b\sqrt{b}$
6) $2xy\sqrt{5xy}$
7) $6x\sqrt[3]{y}$
8) $15x\sqrt{3}$
9) $20x\sqrt{5}$
10) $16y\sqrt[3]{x^2}$
11) $7x\sqrt[3]{xy^2}$
12) $20\sqrt{5a}$

Simplifying Radical Expressions Involving Fractions

1) $\frac{2\sqrt{5mr}}{m^2}$
2) $\frac{8\sqrt{3k}}{k}$
3) $\frac{a-\sqrt{ab}}{a-b}$
4) $\frac{3-\sqrt{5}+3\sqrt{2}-\sqrt{10}}{4}$
5) $\frac{12+2\sqrt{3}+6\sqrt{5}+\sqrt{15}}{33}$
6) $-3+\sqrt{5}$
7) -1
8) $\frac{\sqrt{35}-\sqrt{14}+5\sqrt{10}}{3}$
9) $\frac{24-3\sqrt{10}-4\sqrt{14}+\sqrt{35}}{27}$
10) $\frac{3\sqrt{6}+4}{2}$
11) $4a^2\sqrt{b}$
12) $6x$

Multiplying radical expressions

1) $180\sqrt{55}$
2) $6\sqrt{3}+6$
3) $23x\sqrt{x}$
7) $-12x^2\sqrt{35}$
8) $-90x^2\sqrt{3x}$
9) $-12\sqrt{3}-24$

4) 30

5) $20x^2$

6) $2x\sqrt{6x}$

10) $6\sqrt{2x} - 6x\sqrt{3}$

11) $54x^2$

12) $5\sqrt{15r} + 3\sqrt{5r}$

Adding and Subtracting Radical Expressions

1) $9\sqrt{10}$

2) $-3\sqrt{6}$

3) $-7\sqrt{22}$

4) $2\sqrt{7}$

5) $2\sqrt{3}$

6) $\sqrt{3}$

7) 0

8) $9\sqrt{2}$

9) $4\sqrt{5}$

10) $18\sqrt{2}$

11) $2\sqrt{35}$

12) $-12\sqrt{19}$

Domain and Range of Radical Functions

1) domain: $x \geq -2$
   range: $y \geq -3$

2) domain: {all real numbers}
   range: {all real numbers}

3) domain: $x \geq 2$
   range: $y \geq 5$

4) domain: {all real numbers}
   range: {all real numbers}

5)

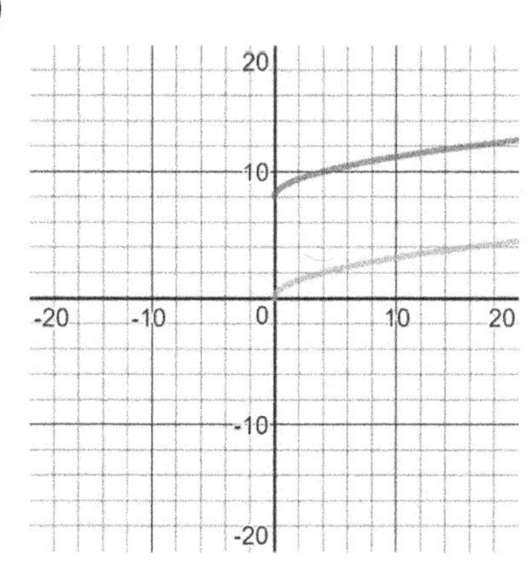

6)

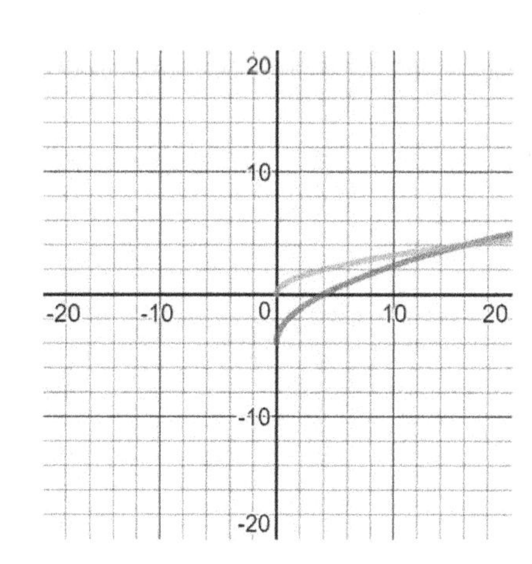

Radical Equations

1) $\{70\}$
2) $\{28\}$
3) $\{16\}$
4) $\{56\}$
5) $\{\sqrt{3}\}$
6) $\{6\}$
7) $\{0\}$
8) $\{\frac{5}{3}\}$
9) $\{61\}$
10) $\{1\}$
11) $\{-2\}$
12) $\{6\}$

Chapter 11:
Rational Expressions

Topics that you'll learn in this chapter:

- ✓ Simplifying Rational Expressions
- ✓ Graphing Rational Expressions
- ✓ Multiplying Rational Expressions
- ✓ Dividing Rational Expressions
- ✓ Adding and Subtracting Rational Expressions
- ✓ Rational Equations
- ✓ Simplify Complex Fractions

Simplifying Rational Expressions

Step-by-step guide:

- ✓ Factorize numerator and denominator if they are factorable.
- ✓ Find common factors of both numerator and denominator.
- ✓ Remove the common factor in both numerator and denominator.
- ✓ Simplify if needed.

Examples:

1) Simplify $\frac{12x^2y}{8y^2}$

   Cancel the common factor 4: $\frac{12x^2y}{8y^2} = \frac{3x^2y}{2y^2}$

   Cancel the common factor y: $\frac{3x^2y}{2y^2} = \frac{3x^2}{2y}$

   Then: $\frac{12x^2y}{8y^2} = \frac{3x^2}{2y}$

2) Simplify $\frac{2x^2 - 2x - 12}{x-3}$

   Factor $2x^2 - 2x - 12 = 2(x+2)(x-3)$

   Then: $\frac{2x^2 - 2x - 12}{x-3} = \frac{2(x+2)(x-3)}{x-3}$

   Cancel the common factor: $(x-3)$

   Then: $\frac{2(x+2)(x-3)}{x-3} = 2(x+2)$

Simplify

1) $\frac{x+3}{3x+9} =$

2) $\frac{16}{4x-4} =$

3) $\frac{36\ \ ^3}{42x^3} =$

4) $\frac{x^2 - 3x - 4}{x^2 + 2x - 24} =$

5) $\frac{81x^3}{18x} =$

6) $\frac{x-3}{x^2 - x - 6} =$

7) $\frac{x^2 - 3x - 28}{x-7} =$

8) $\frac{6x+18}{30} =$

9) $\frac{16}{4x-4} =$

Graphing Rational Expressions

Step-by-step guide:

- ✓ Find the vertical asymptotes of the function, if there is any. (Vertical asymptotes are vertical lines which correspond to the zeroes of the denominator)
- ✓ Find horizontal or slant asymptote. (If numerator has a bigger degree than denominator, there will be slant asymptote.)
- ✓ If denominator has a bigger degree than numerator, the horizontal asymptote is the x-axes or the line y=0. If they have the same degree, the horizontal asymptote equals the leading coefficient (the coefficient of the largest exponent) of the numerator divided by the leading coefficient of the denominator.
- ✓ Find intercepts and plug in some values of x and solve for y and graph

Examples:

Graph rational expressions. $f(x) = \frac{x^2-x+2}{x-2}$

Domain: $\begin{bmatrix} solution: x, 2 \; or \; x > 2 \\ interval \; notation: (-\infty, 2) \cup (2, \infty) \end{bmatrix}$

Range: $\begin{bmatrix} solution: f(x) \leq -1 \; or \; f(x) \geq 7 \\ interval \; notation: (-\infty, -1] \cup [7, \infty) \end{bmatrix}$

Axis interception points of $\frac{x^2-x+2}{x-2}$: y Interceptions: $(0, -1)$

Asymptotes of $\frac{x^2-x+2}{x-2}$: vertical: $x = 2$, horizontal: $y = x + 1$

Extreme points of $\frac{x^2-x+2}{x-2}$: Maximum $(0, -1)$, Minimum $(4, 7)$

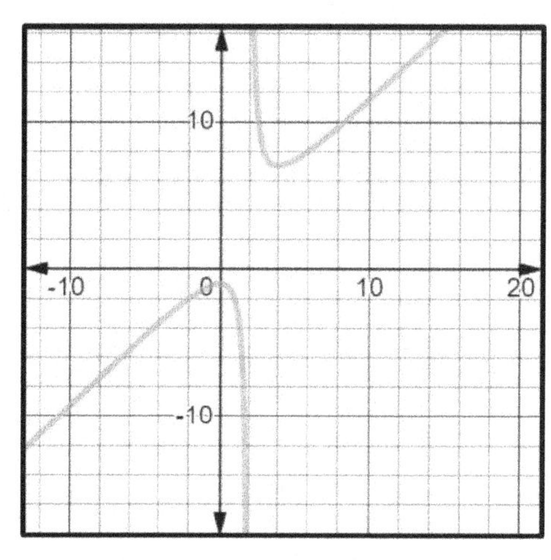

Identify the points of discontinuity, holes, vertical asymptotes, x-intercepts, and horizontal asymptote of each

1) $f(x) = \frac{x^3 - x^2 - 6x}{-3x^3 - 3x + 18}$

2) $f(x) = \frac{x^2 + x - 6}{-4x^2 - 16 \quad - 12}$

3) $f(x) = \frac{x - 2}{x - 4}$

4) $f(x) = \frac{1}{3x^2 + 3x - 18}$

Multiplying Rational Expressions

Step-by-step guide:

✓ Multiplying rational expressions is the same as multiplying fractions. First, multiply numerators and then multiply denominators. Then, simplify as needed.

Examples:

1) **Solve** $\frac{5x+50}{x+10} . \frac{x-2}{5} =$

$\frac{5x+50}{x+10} = \frac{5(x+10)}{x+10} = 5$

then: $\frac{5x+50}{x+10} . \frac{x-2}{5} = \frac{5(x-2)}{5} = x-2$

2) **Solve** $\frac{x-7}{x+6} \times \frac{10x+60}{x-7} =$

Multiply fractions: $\frac{x-7}{x+6} \times \frac{10x+60}{x-7} = \frac{(x-7)(10x+60)}{(x+6)(x-7)}$

Cancel the common factor: $\frac{(x-7)(10x+60)}{(x+6)(x-7)} = \frac{(10 \quad +60)}{(x+6)}$

Factor $10x+60 = 10(x+6)$

Then: $\frac{(10x+60)}{(x+6)} = \frac{10(x+6)}{(x+6)} = 10$

Simplify each expression

1) $\frac{1}{x+10} \times \frac{10 \quad +30}{x+3} =$

2) $\frac{8\,(x+1)}{7x} \times \frac{9}{8\,(x+1)} =$

3) $\frac{2\,(x+6)}{4} \times \frac{x-3}{2\,(x-1)} =$

4) $\frac{9\,(x+4)}{x+4} \times \frac{9x}{9\,(x-5)} =$

5) $\frac{3x^2+18x}{x+6} \times \frac{1}{x+8} =$

6) $\frac{21x^2-21}{18x^2-18} \times \frac{6x}{6x^2} =$

7) $\frac{1}{x-9} \times \frac{x^2+6x-27}{x+9} =$

8) $\frac{x^2-10x+25}{10x-100} \times \frac{x-10}{45-9x} =$

Dividing Rational Expressions

Step-by-step guide:

- ✓ To divide rational expression, use the same method we use for dividing fractions.
- ✓ Keep, Change, Flip
- ✓ Keep first rational expression, change division sign to multiplication, and flip the numerator and denominator of the second rational expression. Then, multiply numerators and multiply denominators. Simplify as needed.

Examples:

1) Solve $\frac{12x}{3} \div \frac{5}{8} =$

$\frac{12x}{3} \div \frac{5}{8} = \frac{\frac{12x}{3}}{\frac{5}{8}}$, Use Divide fractions rules: $\frac{\frac{a}{b}}{\frac{c}{d}} = \frac{a \,.\, d}{b \,.\, c}$

$\frac{\frac{12x}{3}}{\frac{5}{8}} = \frac{12x \,.\, 8}{3 \,.\, 5} = \frac{96x}{15}$

2) Solve $\frac{9a}{a+5} \div \frac{9a}{2a+10} =$

$\frac{\frac{9a}{a+5}}{\frac{9a}{2a+10}}$, Use Divide fractions rules: $\frac{(9a)(a+5)}{(9a)(2a+10)}$

Cancel common fraction: $\frac{(9a)(2a+10)}{(9a)(a+5)} = \frac{(2a+10)}{(a+5)} = \frac{2(a+5)}{(a+5)} = 2$

Divide.

1) $\frac{10\ \ ^{2}}{7} \div \frac{3n}{12} =$

2) $\frac{11x}{x-7} \div \frac{11x}{12\ \ -84} =$

3) $\frac{x+10}{9x^2-90} \div \frac{1}{9x} =$

4) $\frac{x-2}{x+\ 6x-16} \div \frac{11}{x+9} =$

5) $\frac{3x}{x-5} \div \frac{3x}{10\ \ -50} =$

6) $\frac{x+5}{x+13x+40} \div \frac{4x}{x+9} =$

7) $\frac{x+4}{x+\ 14\ \ +40} \div \frac{6x}{x+9} =$

8) $\frac{14\ \ +12}{3} \div \frac{63\ \ +54}{3x} =$

Adding and Subtracting Rational Expressions

Step-by-step guide:

- ✓ For adding and subtracting rational expressions:
- ✓ Find least common denominator (LCD).
- ✓ Write each expression using the LCD.
- ✓ Add or subtract the numerators.
- ✓ Simplify as needed.

Examples:

1) Solve $\frac{2}{6x+10}+\frac{x-6}{6x+10}=$

Use this rule: $\frac{a}{c}\pm\frac{b}{c}=\frac{a\pm b}{c}\rightarrow\frac{2}{6x+10}+\frac{x-6}{6x+10}=\frac{2+x-6}{6x+1}=\frac{x-4}{6x+10}$

2) Solve $\frac{x+2}{x-4}+\frac{x-2}{x+3}=$

Least common multiplier of $(x-4)$ and $(x+3)$: $(x-4)(x+3)$

Then: $\frac{(x+2)(x+3)}{(x-4)(x+3)}+\frac{(x-2)(x-4)}{(x+3)(x-4)}=\frac{(x+2)(x+3)+(x-2)(x-4)}{(x+3)(x-4)}$

Expand: $(x+2)(x+3)+(x-2)(x-4)=2x^2-x+14$

Then: $\frac{2x^2-x+14}{(x+3)(x-4)}$

Simplify each expression.

1) $\frac{3}{x+7}-\frac{4}{x-8}=$

2) $\frac{x-7}{x^2-16}-\frac{x-1}{16-x^2}=$

3) $\frac{5}{x+5}+\frac{4x}{2x+6}=$

4) $2+\frac{x-3}{x+1}=$

5) $\frac{2x}{5x+4}+\frac{6x}{2x+3}=$

6) $\frac{5xy}{x^2-y^2}-\frac{x-y}{x+y}=$

7) $\frac{2}{x^2-5x+4}+\frac{-2}{x^2-4}=$

8) $\frac{4}{x+1}-\frac{2}{x+2}=$

Rational Equations

Step-by-step guide:

- ✓ For solving rational equations, we can use following methods:
- ✓ Converting to a common denominator: In this method, you need to get a common denominator for both sides of equation Then make numerators equal and solve for the variable.
- ✓ Cross-multiplying: This method is useful when there is only one fraction on each side of the equation. Simply multiply first numerator by second denominator and make the result equal to the product of second numerator and first denominator.

Examples: **Solve** $\frac{2x-3}{x+1}=\frac{x+6}{x-2}$

Use fraction cross multiply: if $\frac{a}{b}=\frac{c}{d}$ then: $a.d=b.c$

$\frac{2x-3}{x+1}=\frac{x+6}{x-2}\rightarrow(2x-3)(x-2)=(x+1)(x+6)$

solve: $(2x-3)(x-2)=(x+1)(x+6)\rightarrow 2x^2-7x+6=x^2+7x+6$

$2x^2-7x+6=x^2+7x+6\rightarrow$

Subtract 6 from both sides: $2x^2-7x+6-6=x^2+7x+6-6$

Simplify: $2x^2-7x=x^2+7x\rightarrow$ subtract $7x$ from both sides: $2x^2-14x=x^2$

Subtract x^2 both sides: $x^2-14x=0$

Now solve with quadratic equation formula: $x_{1,2}=\frac{-b\pm\sqrt{b^2-4a}}{2a}$

$x_1=\frac{-(-14)+\sqrt{(-14)^2-4.1.0}}{2.1}=14$

$x_2=\frac{-(-14)-\sqrt{(-14)^2-4.1.0}}{2.1}=0$

Solve each equation. Remember to check for extraneous solutions.

1) $\frac{2x-3}{x+1}=\frac{x+6}{x-2}$

2) $\frac{1}{6b^2}+\frac{1}{6b}=\frac{1}{b^2}$

3) $\frac{3x-2}{9x+1}=\frac{2x-5}{6x-5}$

4) $\frac{1}{n^2}+\frac{1}{n}=\frac{1}{2n^2}$

5) $\frac{1}{8b^2}=\frac{1}{4b^2}-\frac{1}{b}$

6) $\frac{1}{n-8}-1=\frac{7}{n-8}$

7) $\frac{5}{r-2}=-\frac{10}{r+2}+7$

8) $1=\frac{1}{x^2+2x}+\frac{x-1}{x}$

Simplify Complex Fractions

- ✓ Convert mixed numbers to improper fractions.
- ✓ Simplify all fractions.
- ✓ Write the fraction in the numerator of the main fraction line then write division sing (÷) and the fraction of the denominator.
- ✓ Use normal method for dividing fractions.
- ✓ Simplify as needed.

Examples: **Solve** $\frac{\frac{4}{5}}{\frac{2}{25}-\frac{5}{16}}$

Use the fraction rule: $\frac{\frac{b}{c}}{a}=\frac{b}{c\,.\,a}$

$$\frac{\frac{4}{5}}{\frac{2}{25}-\frac{5}{16}}=\frac{4}{5(\frac{2}{25}-\frac{5}{16})}=\frac{4}{5(-\frac{93}{400})}=\frac{4}{-5.\frac{93}{400}}=-\frac{4}{5.\frac{93}{400}}=-\frac{4}{\frac{93}{80}}$$

Use the fraction rule: $\frac{a}{\frac{b}{c}}=\frac{a\,.\,c}{b}\rightarrow-\frac{4}{\frac{93}{80}}=-\frac{4\,.\,80}{93}=-\frac{320}{93}$

Simplify each expression.

1) $\frac{\frac{14}{3}}{-6\frac{2}{11}}=$

2) $\frac{9}{\frac{9}{x}+\frac{2}{3x}}=$

3) $\frac{x^2}{\frac{4}{5}-\frac{4}{x}}=$

4) $\frac{\frac{4}{x-3}-\frac{2}{x+2}}{\frac{8}{x^2+6x+8}}=$

5) $\frac{\frac{16}{x-1}}{\frac{16}{5}-\frac{16}{25}}=$

6) $\frac{2+\frac{6}{x-4}}{2-\frac{4}{x-4}}=$

7) $\frac{\frac{1}{2}-\frac{x+5}{4}}{\frac{x^2}{2}-\frac{5}{2}}=$

8) $\frac{\frac{x-6}{2}-\frac{x-2}{x-6}}{\frac{36}{x-2}+\frac{4}{9}}=$

Answers of Worksheets – Chapter 11

Simplifying rational expressions

1) $\frac{1}{3}$
2) $\frac{4}{x-1}$
3) $\frac{6}{7}$
4) $\frac{x+1}{x+6}$
5) $\frac{9x^2}{2}$
6) $\frac{x+3}{5}$
7) $x+4$
8) $\frac{x+3}{8}$
9) $\frac{4}{x-1}$

Graphing rational expressions

1) Discontinuities: $-3, 2$
   Vertical Asym: $x = -3, x = 2$
   Holes: None
   Horz. Asym: None
   x–intercepts: $0, -2, 3$

2) Discontinuities: $-1, -3$
   Vertical Asym: $x = -1$
   Holes: $x = -3$
   Horz. Asym: $y = -\frac{1}{4}$
   x–intercepts: 2

3) Discontinuities: 4
   Vertical Asym: $x = 4$
   Holes: None
   Horz. Asym: $y = 1$
   x–intercepts: 2

4) Discontinuities: $-3, 2$
   Vertical Asym: $x = -3, x = 2$
   Holes: None
   Horz. Asym: $y = 0$
   x–intercepts: None

5)

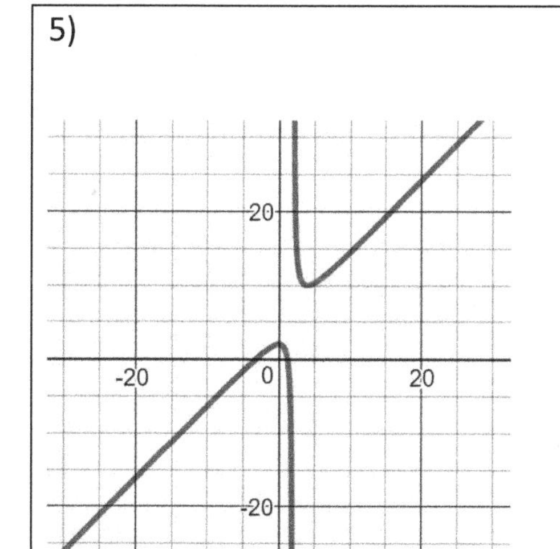

6)

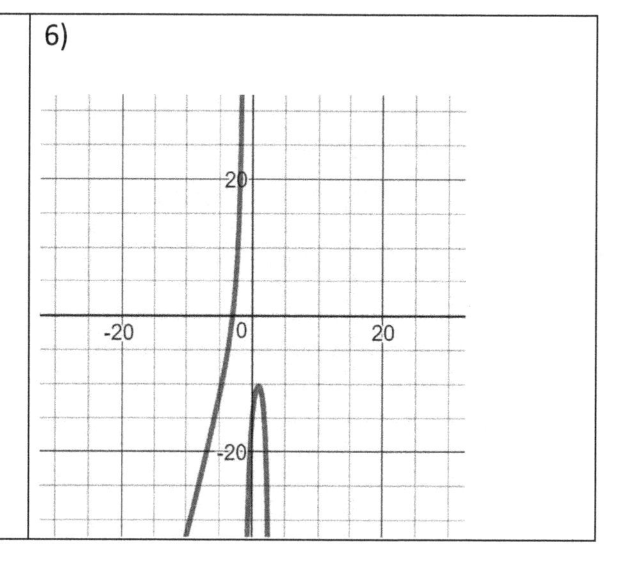

Multiplying rational expressions

1) $\frac{10}{x+10}$
2) $\frac{9}{7x}$
3) $\frac{x+6}{4}$
4) $\frac{9x}{x-5}$
5) $\frac{3x}{x+8}$
6) $\frac{7}{6x}$
7) $\frac{x-3}{x-9}$
8) $-\frac{(x-5)}{90}$

Dividing rational expressions

1) $\frac{40n}{7}$
2) 12
3) $\frac{x+10}{x-10}$
4) $\frac{x+9}{11\ (x+8)}$
5) 10
6) $\frac{2x}{9}$
7) $\frac{x+9}{6x\ (x+10)}$
8) $\frac{2x}{9}$

Adding and subtracting rational expressions

1) $\frac{7x+4}{(x+7)(x-8)}$
2) $\frac{2}{x+4}$
3) $\frac{x-5}{x+2}$
4) $\frac{3x-1}{x+1}$
5) $\frac{34x^2+30x}{(5x+4)(2x+3)}$
6) $\frac{-x^2+7xy-y^2}{(x-y)(x+y)}$
7) $\frac{10x-16}{(x^2-5x+4)(x^2-4)}$
8) $\frac{2x+6}{(x+1)(x+2)}$

Solving Rational Equations

1) $\{0, 14\}$
2) $\{-\frac{15}{16}\}$
3) $\{\frac{1}{6}\}$
4) $\{-\frac{1}{2}\}$
5) $\{\frac{1}{8}\}$
6) $\{2\}$
7) $\{-\frac{6}{7}, 3\}$
8) $\{-1\}$

Simplify complex fractions

1) $-\frac{77}{102}$

2) $\frac{27}{29}$

3) $\frac{5x^2}{4x-20}$

4) $\frac{(x+7)(x+4)}{4\,(x-3)}$

5) $\frac{25}{4x-4}$

6) $\frac{x-1}{x-6}$

7) $\frac{-3-x}{2x^2-10}$

8) $\frac{3x^3-60x^2+252x-288-x}{584x+8x^2-3792}$

Chapter 12:
Conic Sections

Topics that you'll learn in this chapter:

- ✓ Equation of a Parabola
- ✓ Finding the Focus, Vertex, and Directrix of a Parabola
- ✓ Standard Form of a Circle
- ✓ Finding the Center and the Radius of Circles
- ✓ Equation of Each Ellipse and Finding the Foci, Vertices, and Co– Vertices of Ellipses
- ✓ Hyperbola in Standard Form and Vertices, Co– Vertices, Foci, and Asymptotes of a Hyperbola
- ✓ Classifying a Conic Section (in Standard Form)

Equation of a Parabola

Step-by-step guide:

✓ The standard form of a parabola:

When it opens up or down:

$(x-h)^2 = 4p(y-k)$, Vertex: (h,k), Directrix: $y = k - p$, Focus: $(h, k+p)$

When it opens right or left:

$(y+k)^2 = 4p(x-h)$, Vertex: (h,k), Directrix: $x = h - p$, Focus: $(h+p, k)$

Examples:

1) *Write the equation of the parabola with* Vertex (0, 0) and Focus (0, 2).

The standard form of a parabola: $(x-h)^2 = 4p(y-k)$

Vertex: $(h,k) = (0,0)$, then: $h = 0$, $k = 0$

Focus: $(h, k+p) = (0,2)$, then: $p = 2$

$(x-0)^2 = 4(2)(y-0)$, then: $x^2 = 8y$

2) *Write the equation of the parabola with* Vertex (3, 2) and Focus (3, 4).

The standard form of a parabola: $(x-h)^2 = 4p(y-k)$

Vertex: $(h,k) = (3,2)$, then: $h = 3$, $k = 2$

Focus: $(h, k+p) = (3,4)$, then: $k + p = 4 \rightarrow 2 + p = 4 \rightarrow p = 2$

$(x-3)^2 = 4(2)(y-2)$, then: $(x-3)^2 = 8(y-2)$

Write the equation of the following parabolas.

1) Vertex $(1,1)$ and Focus $(1,6)$
2) Vertex $(-1, 2)$ and Focus $(-1,5)$
3) Vertex $(2,2)$ and Focus $(2,6)$
4) Vertex $(0,1)$ and Focus $(0,2)$
5) Vertex $(2,1)$ and Focus $(4,1)$
6) Vertex $(5,0)$ and Focus $(9,0)$
7) Vertex $(-2,4)$ and Focus $(2,4)$
8) Vertex $(-4, 2)$ and Focus $(0, 2)$

Finding the Focus, Vertex, and Directrix of a Parabola

Step-by-step guide:

✓ The standard form of a parabola:

When it opens up or down:

$(x-h)^2 = 4p(y-k)$, Vertex: (h,k), Directrix: $y = k - p$, Focus: $(h, k+p)$

When it opens right or left:

$(y+k)^2 = 4p(x-h)$, Vertex: (h,k), Directrix: $x = h - p$, Focus: $(h+p, k)$

✓ The vertex of an up-down facing parabola of the form $y = ax^2 + bx + c$ is:

$$x_v = -\frac{b}{2a}$$

Examples:

1) *Write the vertex form equation of parabola.* $y = x^2 + 8x$

The parabola params are: $a = 1, b = 8$ and $c = 0$

$x_v = -\frac{b}{2a} \rightarrow x_v = -\frac{8}{2(1)} = -4$

Use x_v to $find$ y_v: $y_v = ax + bx + c \rightarrow y_v = (1)(-4)^2 + (8)(-4) + 0 \rightarrow y_v = -16$

Therefore, the parabola vertex is: $(-4, -16)$

2) *Write the vertex form equation of parabola.* $y = x^2 - 6x + 5$

$(x-h)^2 = 4p(y-k)$ is the standard equation for an up-down facing parabola with vertex at (h,k), and a focal length p. Rewrite $y = x^2 - 6x + 5$ in standard form:

$4.\frac{1}{4}(y-(-4)) = (x-3)^2$, $(h,k) = (3,-4), p = \frac{1}{4}$

Parabola is symmetric around the y-axis and so the focus lies a distance p from the center $(3,-4)$ along the y-axis: $(3, -4+p) = (3, -4+\frac{1}{4}) = (3, -\frac{15}{4})$

✍ ***Use the information provided to write the vertex form equation of each parabola.***

1) $162 + 731 = -y - 9x^2$
2) $y = x^2 + 16x + 71$
3) Focus: $(-\frac{63}{8}, -7)$, Directrix: $x = -\frac{65}{8}$
4) Focus: $(\frac{107}{12}, -7)$, Directrix: $x = \frac{109}{12}$

Standard Form of a Circle

Step-by-step guide:

- ✓ Equation of circles in standard form: $(x-h)^2 + (y-k)^2 = r^2$
- ✓ Center: (h, k), Radius: r
- ✓ General format: $ax^2 + by^2 + cx + dy + e = 0$

Examples: Write the standard form equation of each circle.

1) $x^2 + y^2 - 8x - 6y + 21 = 0$

$(x-h)^2 + (y-k)^2 = r^2$ is the circle equation with a radius r, centered at (h, k)

First move the loose number to the right side: $x^2 + y^2 - 8x - 6y = -21$

Group x-variables and y-variables together: $(x^2 - 8x) + (y^2 - 6y) = -21$

Convert x to square form:

$(x^2 - 8x + 16) + (y^2 - 6y) = -21 + 16 \rightarrow (x-4)^2 + (y^2 - 6y) = -21 + 16 \rightarrow$

Convert y to square form:

$(x-4)^2 + (y^2 - 6y + 9) = -21 + 16 + 9 \rightarrow (x-4)^2 + (y-3)^2 = 4$

Then: $(x-4)^2 + (y-3)^2 = 2^2$

2) Center: $(-9, -12)$, Radius: 4

$(x-h)^2 + (y-k)^2 = r^2$ is the circle equation with a radius r, centered at (h, k)

$h = -9$, $k = -12$ and $r = 4$

Then: $(x-(-9))^2 + (y-(-12))^2 = (4)^2 \rightarrow (x+9)^2 + (y+12)^2 = 16$

✍ ***Write the standard form equation of each circle.***

1) $y^2 + 2x + x^2 = 24y - 120$
2) $x^2 + y^2 - 2y - 15 = 0$
3) $8x + x^2 - 2y = 64 - y^2$
4) Center: (–5, –6), Radius: 9
5) Center: (–12, –5), Area: 4π
6) Center: (–11, –14), Area: 16π
7) Center: (–3, 2), Circumference: 2π
8) Center: (15, 14), Circumference: $2\pi\sqrt{15}$

Finding the Center and the Radius of Circles

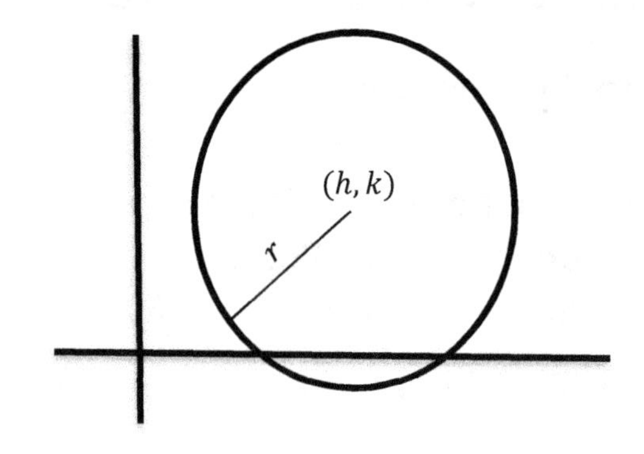

Step-by-step guide:

$(x - h)^2 + (y - k)^2 = r^2$

center: (h, k), radius: r

Examples: Identify the center and radius.

1) $x^2 + y^2 - 4y + 3 = 0$

$(x - h)^2 + (y - k)^2 = r^2$ is the circle equation with a radius r, centered at (h, k)

Rewrite $x^2 + y^2 - 4y + 3 = 0$ in the form of the standard circle equation:

$(x - 0)^2 + (y - 2)^2 = 1^2$

Then: center: $(0,2)$ and $r = 1$

2) $4x + x^2 - 6y = 24 - y^2$

$(x - h)^2 + (y - k)^2 = r^2$ is the circle equation with a radius r, centered at (h, k)

Rewrite $4x + x^2 - 6y = 24 - y^2$ in the form of the standard circle equation:

$(x - (-2))^2 + (y - 3)^2 = \sqrt{37}^2$

Then: center: $(-2,3)$ and $r = \sqrt{37}$

✍ ***Identify the center and radius of each. Then sketch the graph.***

1) $(x - 2)^2 + (y + 5)^2 = 10$

2) $x^2 + (y - 1)^2 = 4$

3) $(x - 2)^2 + (y + 6)^2 = 9$

4) $(x + 14)^2 + (y - 5)^2 = 16$

Equation of Each Ellipse and Finding the Foci, Vertices, and Co–Vertices of Ellipses

Step-by-step guide:

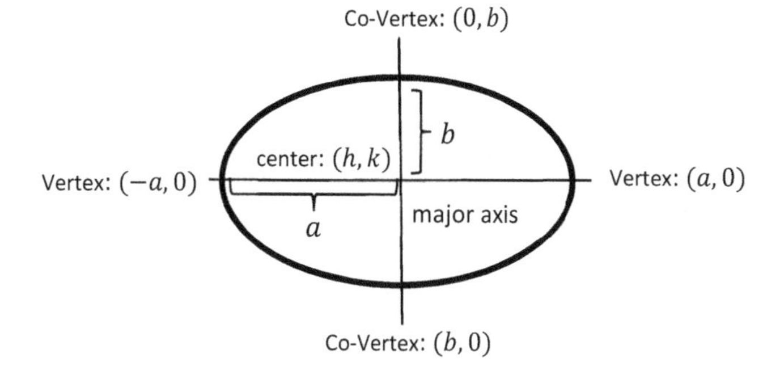

Horizontal: $\frac{(x-h)^2}{a^2} + \frac{(y-k)^2}{b^2} = 1$

Vertical: $\frac{(x-h)^2}{b^2} + \frac{(y-k)^2}{a^2} = 1$

Vertices: the vertices are the two points on the ellipse that intersect the major axis for an ellipse with major axis parallel to the x-axis, the vertices are:

$$(h+a, k), (h-a, k)$$

Foci: $(h+c, k), (h-c, k)$, where $c = \sqrt{a^2 - b^2}$ is the distance from the center (h, k) to a focus

Examples:

1) find vertices of $\frac{x^2}{169} + \frac{y^2}{64} = 1$

Rewrite $\frac{x^2}{169} + \frac{y^2}{64} = 1$ in the form of the standard ellipse equation

$\frac{(x-h)^2}{a^2} + \frac{(y-k)^2}{b^2} = 1 \rightarrow \frac{(x-0)^2}{13^2} + \frac{(y-0)^2}{8^2} = 1$

Then: $(h, k) = (0,0)$, $a = 13$ and $b = 8$

Vertices: $(h+a, k), (h-a, k) \rightarrow (0+13, 0), (0-13, 0) \rightarrow (13,0), (-13,0)$

Identify the vertices, co–vertices, foci.

1) $\frac{x^2}{36} + \frac{y^2}{16} = 1$

2) $\frac{x^2}{49} + \frac{y^2}{169} = 1$

3) $\frac{(x+5)^2}{81} + \frac{(y-1)^2}{144} = 1$

4) $\frac{(x-3)^2}{49} + \frac{(y-9)^2}{4} = 1$

5) $\frac{x^2}{64} + \frac{(y-8)^2}{9} = 1$

6) $\frac{x^2}{64} + \frac{(y-6)^2}{121} = 1$

Hyperbola in Standard Form and Vertices, Co– Vertices, Foci, and Asymptotes of a Hyperbola

Step-by-step guide:

$\frac{(y-k)^2}{a^2} - \frac{(x-h)^2}{b^2} = 1$

center: (h, k)

foci: $(h, k \pm c)$

vertices: $(h, k \pm a)$

transverse axis: $x = h$

(parallel to y-axis)

asymptotes: $y - \mathrm{k} = \pm\frac{a}{\mathrm{b}}(x - \mathrm{h})$

$\frac{(x-h)^2}{a^2} - \frac{(y-k)^2}{b^2} = 1$

center: (h, k)

foci: $(h \pm c, k)$

vertices: $(h \pm a, k)$

transverse axis: $y = k$

(parallel to x-axis)

asymptotes: $y - \mathrm{k} = \pm\frac{b}{\mathrm{a}}(x - \mathrm{h})$

Examples: Find center and foci of $-x^2 + y^2 - 18x - 14y - 132 = 0$

Rewrite in standard form:

Add 132 to both sides: $-x^2 + y^2 - 18x - 14y = 132$

Factor out coefficient of square terms: $-(x^2 + 18) + (y^2 - 14y) = 132$

Convert x and y to square form: $-(x^2 + 18 + 81) + (y^2 - 14y + 49) = 132 - 81 + 49$

$-(x + 9)^2 + (y - 7)^2 = 132 - 81 + 49 \rightarrow -(x + 9)^2 + (y - 7)^2 = 100$

Divide by 100: $-\frac{(x+9)^2}{100} + \frac{(y-7)^2}{100} = 1 \rightarrow \frac{(x-7)^2}{10^2} - \frac{(x-(-9))^2}{10^2} = 1$

Then: $(\mathrm{h}, \mathrm{k}) = (-9, 7)$, $\mathrm{a} = 10$, $\mathrm{b} = 10$ and center is: $(-9, 7)$

foci: $(-9, 7 + \mathrm{c}), (-9, 7 - \mathrm{c})$

Compute c: $\mathrm{c} = \sqrt{10^2 + 10^2} = 10\sqrt{2}$, then: $(-9, 7 + 10\sqrt{2}), (-9, 7 - 10\sqrt{2})$

Identify the vertices, foci, and direction of opening of each.

1) $\frac{y^2}{25} - \frac{x^2}{16} = 1$

2) $\frac{x^2}{121} - \frac{y^2}{36} = 1$

3) $\frac{x^2}{121} - \frac{y^2}{81} = 1$

4) $\frac{x^2}{81} - \frac{y^2}{4} = 1$

5) $\frac{(x+2)^2}{169} - \frac{(y+8)^2}{4} = 1$

6) $\frac{(y+8)^2}{36} - \frac{(y+2)^2}{25} = 1$

Classifying a Conic Section (in Standard Form)

Step -by-step guide:

Conic section	Standard form of equation	
Parabola	$y = a(x-h)^2 + k$	$x = a(y-k)^2 + h$
Circle	$(x-h)^2 + (y-k)^2 = r^2$	
Ellipse	$\frac{(x-h)^2}{a^2} + \frac{(y-k)^2}{b^2} = 1$	$\frac{(y-k)^2}{a^2} + \frac{(x-h)^2}{b^2} = 1$
Hyperbola	$\frac{(x-h)^2}{a^2} - \frac{(y-k)^2}{b^2} = 1$	$\frac{(y-k)^2}{a^2} - \frac{(x-h)^2}{b^2} = 1$

Examples:

Write this equation in standard form. $-x^2 + 10x + y - 21 = 0$

It's parabola. Rewrite in standard form: $-x^2 + 10x + y - 21 = 0$

Rewrite as: $x^2 = y^2 - 8y + 17$, complete the square $y^2 - 8y + 17 = (y-4)^2 + 1$

Then: $x = (y-4)^2 + 1$

Subtract 1 from both sides: $x - 1 = (y-4)^2$

Rewrite in standard form: $4.\frac{1}{4}(x-1) = (y-4)^2$

Write this equation in standard form. $x^2 - 4y^2 + 6x - 8y + 1 = 0$

It's Hyperbola. First subtract 1 from both sides:

$x^2 - 4y^2 + 6x - 8y + 1 = 0 \rightarrow x^2 - 4y^2 + 6x - 8y = -1$

Factor out coefficient of square terms: $(x^2 + 6x) - 4(y^2 + 2y) = -1$

Divide by coefficient of square terms: $\frac{1}{4}(x^2 + 6x) - (y^2 + 2y) = -\frac{1}{4}$

Convert x and y to square form: $\frac{1}{4}(x^2 + 6x + 9) - (y^2 + 2y + 1) = -\frac{1}{4} + \frac{1}{4}(9) - 1$

Convert y to square form: $\frac{1}{4}(x+3)^2 - (y+1)^2 = -\frac{1}{4} + \frac{1}{4}(9) - 1$

Then: $\frac{1}{4}(x+3)^2 - (y+1)^2 = 1 \rightarrow \frac{(x+3)^2}{4} - \frac{(y+1)^2}{1} = 1 \rightarrow \frac{(x-(-3))^2}{2^2} - \frac{(y-(-1))^2}{1^2} = 1$

Classify each conic section and write its equation in standard form.

1) $3x^2 + 3x + y + 79 = 0$

2) $x^2 + y^2 + 4x - 2y - 18 = 0$

3) $49x^2 + 9y^2 + 392x + 343 = 0$

4) $-9x^2 + y^2 - 72x - 153 = 0$

5) $-2y^2 + x - 20y - 49 = 0$

6) $-x^2 + 10x + y - 21 = 0$

Answers of Worksheets – Chapter 12

Finding the Equation of a Parabola

1) Vertex $(1,1)$ and Focus $(1,6)$: $(x-1)^2 = 20\ (y-1)$
2) Vertex $(-1,2)$ and Focus $(-1,5)$: $(x+1)^2 = 12\ (y-2)$
3) Vertex $(2,2)$ and Focus $(2,6)$: $(x-2)^2 = 8\ (y-2)$
4) Vertex $(0,1)$ and Focus $(0,2)$: $x^2 = 8\ (y-1)$
5) Vertex $(2,1)$ and Focus $(4,1)$: $(y-1)^2 = 8\ (x-2)$
6) Vertex $(5,0)$ and Focus $(9,0)$: $(y-1)^2 = 8\ (x-2)$
7) Vertex $(-2,4)$ and Focus $(2,4)$: $(y-4)^2 = 16(x+2)$
8) Vertex $(-4,2)$ and Focus $(0,2)$: $(y+4)^2 = 16x$

Finding the Focus, Vertex, and Directrix of a Parabola

1) $y = -9\ (x+9)^2 - 2$
2) $y = (x+8)^2 + 7$
3) $x = 2(y+7)^2 - 8$
4) $x = -3(y+7)^2 + 9$

Standard Form of a Circle

1) $(x+1)^2 + (y-12)^2 = 25$
2) $x^2 + (y-1)^2 = 16$
3) $(x+4)^2 + (y-1)^2 = 81$
4) $(x+5)^2 + (y+6)^2 = 81$
5) $(x+12)^2 + (y+5)^2 = 4$
6) $(x+11)^2 + (y+14)^2 = 16$
7) $(x+3)^2 + (y-2)^2 = 1$
8) $(x-15)^2 + (y-14)^2 = 15$

Finding the Center and the Radius of Circles

1) $Center$: $(2,-5)$, $Radius$: $\sqrt{10}$
2) $Center$: $(0,1)$, $Radius$: $2\sqrt{26}$
3) $Center$: $(2,-6)$, $Radius$: 3
4) $Center$: $(-14,-5)$, $Radius$: 4

Equation of Each Ellipse and Finding the Foci, Vertices, and Co–Vertices of Ellipses

1) Vertices: $(6,0), (-6,0)$

Co–vertices: $(0,4), (0,-4)$

Foci: $(2\sqrt{5},0), (-2\sqrt{5},0)$

2) Vertices: $(0,13), (0,-13)$

Co–vertices: $(7,0), (-7,0)$

Foci: $(0,2\sqrt{30}), (0,-2\sqrt{30})$

3) Vertices: $(-5, 13), (-5, -11)$

Co–vertices: $(4, 1), (-14, 1)$

Foci: $(-5, 1+3\sqrt{7}), (-5, 1-3\sqrt{7})$

4) Vertices: $(10, 9), (-4, 9)$

Co–vertices: $(3, 11), (3, 7)$

Foci: $(3+3\sqrt{5}, 9), (3-3\sqrt{5}, 9)$

5) Vertices: $(8, 8), (-8, 8)$

Co–vertices: $(0, 11), (0, 5)$

Foci: $(\sqrt{55}, 8), (-\sqrt{55}, 8)$

6) Vertices: $(0, 17), (0, -5)$

Co–vertices: $(8, 6), (-8, 6)$

Foci: $(0, 6+\sqrt{57}), (0, 6-\sqrt{57})$

Hyperbola in Standard Form and Vertices, Co– Vertices, Foci, and Asymptotes of a Hyperbola

1) Vertices: $(0, 5), (0, -5)$

   Foci: $(0, \sqrt{41}), (0, -\sqrt{41})$

   Opens up/down

2) Vertices: $(11, 0), (-11, 0)$

   Foci: $(\sqrt{157}, 0), (-\sqrt{157}, 0)$

   Opens left/right

3) Vertices: $(11, 0), (-11, 0)$

   Foci: $(\sqrt{202}, 0), (-\sqrt{202}, 0)$

   Opens left/right

4) Vertices: $(9, 0), (-9, 0)$

   Foci: $(\sqrt{85}, 0), (-\sqrt{85}, 0)$

   Opens left/right

5) Vertices: $(11, -8), (-15, -8)$

   Foci:

   $(-2+\sqrt{173}, -8), (-2-\sqrt{173}, -8)$

   Opens left/right

6) Vertices: $(-2, -2), (-2, -14)$

   Foci: $(-2, -8+\sqrt{61}), (-2, -8-\sqrt{61})$

   Opens up/down

Classifying a Conic Section (in Standard Form)

1) Parabola, $y = -3(x+5)^2 - 4$
2) Circle, $(x+2)^2 + (y-1)^2 = 23$
3) Ellipse, $\frac{(x+4)^2}{9} + \frac{y^2}{49} = 1$
4) Hyperbola, $\frac{y^2}{9} - (x+4)^2 = 1$
5) Parabola, $x = 2(y+5)^2 - 1$
6) Parabola, $y = (x-5)^2 - 4$

Chapter 13: Trigonometric Functions

Topics that you'll learn in this chapter:

- ✓ Trig Ratios of General Angles
- ✓ Conterminal Angles and Reference Angles
- ✓ Angles and Angle Measure
- ✓ Evaluating Trigonometric Function
- ✓ Missing Sides and Angles of a Right Triangle
- ✓ Arc Length and Sector Area

Trig Ratios of General Angles

Step-by-step guide:

✓ Learn common trigonometric functions:

θ	$0°$	$30°$	$45°$	$60°$	$90°$
$\sin\theta$	0	$\frac{1}{2}$	$\frac{\sqrt{2}}{2}$	$\frac{\sqrt{3}}{2}$	1
$\cos\theta$	1	$\frac{\sqrt{3}}{2}$	$\frac{\sqrt{2}}{2}$	$\frac{1}{2}$	0
$\tan\theta$	0	$\frac{\sqrt{3}}{3}$	1	$\sqrt{3}$	Undefined

Examples:

Find each trigonometric function.

1) $sin - 120°$. Use the following property: $sin(-x) = -sin(x)$

$sin - 120° = -sin\ 120°$. $sin\ 120° = \frac{\sqrt{3}}{2}$, then: $sin - 120° = -\frac{\sqrt{3}}{2}$

2) $cos\ 150°$

Recall that $cos\ 150° = -cos\ 30°$. Then: $cos\ 150° = -cos\ 30° = -\frac{\sqrt{3}}{2}$

Evaluate.

1) $sin -60° =$ ________
2) $sin\ 150° =$ ________
3) $cos\ 315° =$ ________
4) $cos\ 180° =$ ________
5) $sin\ 120° =$ ________
6) $sin -330° =$ ________
7) $tan -90° =$ ________
8) $cot\ 90° =$ ________
9) $tan\ 270° =$ ________
10) $cot\ 150° =$ ________
11) $sec\ 120° =$ ________
12) $csc -360° =$ ________

Conterminal Angles and Reference Angles

Step-by-step guide:

- ✓ Conterminal angles are equal angles.
- ✓ To find a conterminal of an angle, add or subtract 360 degrees (or 2π for radians) to the given angle.
- ✓ Reference angle is the smallest angle that you can make from the terminal side of an angle with the x-axis.

Examples:

1) Find a positive and a negative conterminal angles to angle $70°$.

$70° - 360° = -290°$

$70° + 360° = 430°$

$-290°$ and a $430°$ are conterminal with a $70°$.

2) Find a positive and negative conterminal angles to angle $\frac{\pi}{4}$.

$\frac{\pi}{4} + 2\pi = \frac{9\pi}{4}$

$\frac{\pi}{4} - 2\pi = -\frac{7\pi}{4}$

Find a conterminal angle between 0° and 360° for each angle provided.

1) $-440° =$

2) $640° =$

3) $-435° =$

4) $-330° =$

Find a conterminal angle between 0 and 2π for each given angle.

5) $\frac{15}{4} =$

6) $-\frac{19\pi}{12} =$

7) $-\frac{35\pi}{18} =$

8) $\frac{11}{3} =$

Angles and Angle Measure

Step-by-step guide:

- ✓ To convert degrees to radians, use this formula: $\boldsymbol{Radians} = \boldsymbol{Degrees} \times \frac{\pi}{180}$
- ✓ To convert radians to degrees, use this formula: $\boldsymbol{Degrees} = \boldsymbol{Radians} \times \frac{180}{\pi}$

Examples:

1) Convert 150 degrees to radians.

Use this formula: $Radians = Degrees \times \frac{\pi}{180}$

$$Radians = 150 \times \frac{\pi}{180} = \frac{150\pi}{180} = \frac{5\pi}{6}$$

2) Convert $\frac{2\pi}{3}$ to degrees.

$\frac{2\pi}{3} \times \frac{180}{\pi} = \frac{360\pi}{3\pi} = 120$

Use this formula: $Degrees = Radians \times \frac{\pi}{180}$

$$Radians = \frac{2\pi}{3} \times \frac{180}{\pi} = \frac{360\pi}{3\pi} = 120$$

Convert each degree measure into radians.

1) $-140° =$ ____
2) $320° =$ ____
3) $210° =$ ____
4) $780° =$ ____
5) $-190° =$ ____
6) $345° =$ ____

Convert each radian measure into degrees.

7) $\frac{\pi}{30} =$
8) $\frac{4\pi}{5} =$
9) $\frac{7\pi}{18} =$
10) $\frac{\pi}{5} =$
11) $-\frac{5\pi}{4} =$
12) $\frac{14\pi}{3} =$

Evaluating Trigonometric Function

Step-by-step guide:

- ✓ Step 1: Draw the terminal side of the angle.
- ✓ Step 2: Find reference angle. (It is the smallest angle that you can make from the terminal side of an angle with the x-axis.)
- ✓ Step 3: Find the trigonometric function of the reference angle.

Examples:

1) ***Find the exact value of trigonometric function.*** $\cos 225°$

Write $\cos(225°)$ as $\cos(180° + 45°)$. Recall that $\cos 180° = -1, \cos 45° = \frac{\sqrt{2}}{2}$

225° is in the third quadrant and cosine is negative in the quadrant 3. T***he reference angle of*** 225° is 45°. Therefore, $\cos 225° = -\frac{\sqrt{2}}{2}$

2) ***Find the exact value of trigonometric function.*** $\tan \frac{7\pi}{6}$

Rewrite the angles for $n\ \frac{7\pi}{6}$:

$\tan \frac{7\pi}{6} = \tan\left(\frac{6\pi+\pi}{6}\right) = \tan(\pi + \frac{1}{6}\pi)$

Use the periodicity of tan: $\tan(x + \pi . k) = \tan(x)$

$\tan\left(\pi + \frac{1}{6}\pi\right) = \tan\left(\frac{1}{6}\pi\right) = \frac{\sqrt{3}}{3}$

Find the exact value of each trigonometric function.

1) $\tan -\frac{\pi}{6} =$ ______

2) $\cot -\frac{7\pi}{6} =$ ______

3) $\cos -\frac{\pi}{4} =$ ______

4) $\cos -480° =$ ______

5) $\sin 690° =$ ______

6) $\tan 420° =$ ______

Missing Sides and Angles of a Right Triangle

Step-by-step guide:

- ✓ By using Sine, Cosine or Tangent, we can find an unknown side in a right triangle when we have one length, and one angle (apart from the right angle).
- ✓ Adjacent, Opposite and Hypotenuse, in a right triangle is shown below.
- ✓ Recall the three main trigonometric functions:

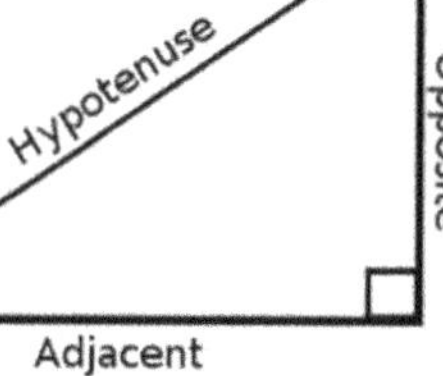

SOH – CAH – TOA, $sine\ \theta = \frac{opposite}{hypotenus}$, $Cos\ \theta = \frac{adjacent}{hypotenuse}$, $\tan\theta = \frac{opposite}{adjacent}$

Example:

Find AC in the following triangle. Round answers to the nearest tenth.

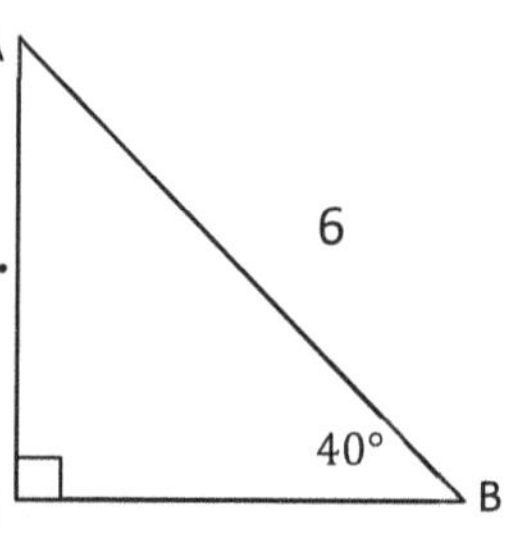

$sine\ \theta = \frac{opposite}{hypotenus}$. $sine\ 40° = \frac{AC}{6} \rightarrow 6\ \times sine\ 40° = AC,$

now use a calculator to find $sine\ 40°$. $sine\ 40° \cong 0.642 \rightarrow AC \cong 3.9$

Find the value of each trigonometric ratio as fractions in their simplest form.

1) $tan\ A$

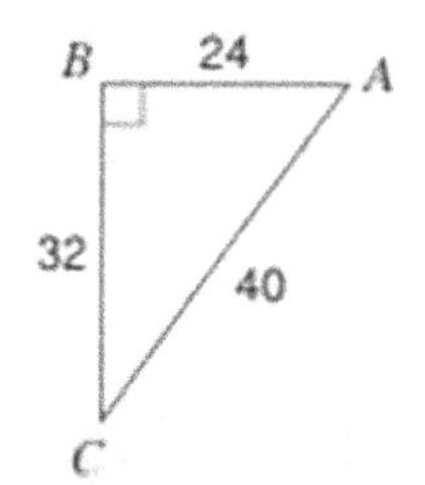

2) $sin\ x$

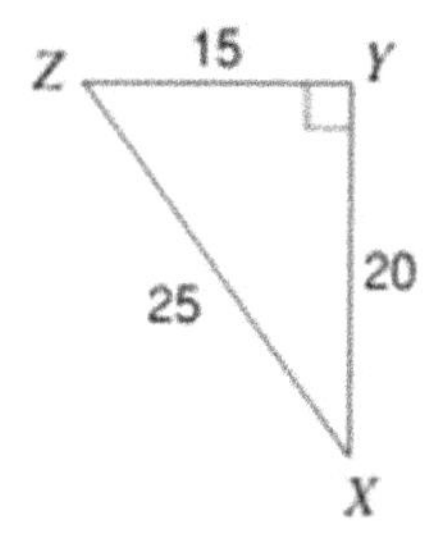

Find the missing sides. Round answers to the nearest tenth.

3)

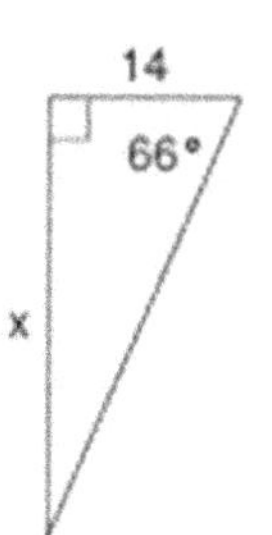

4)

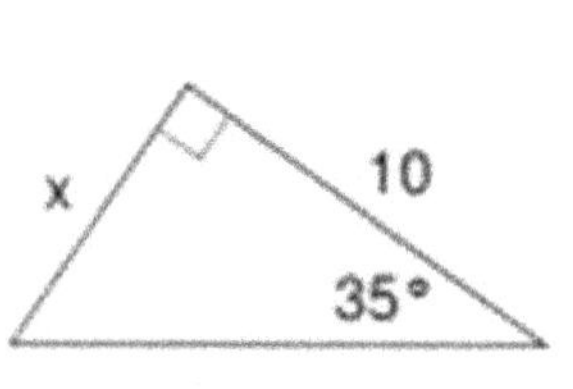

5)

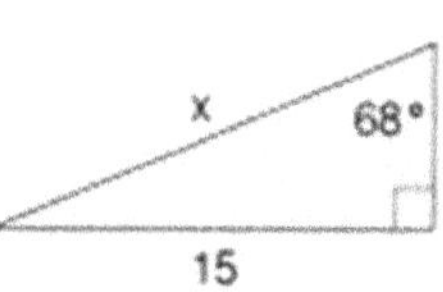

6)

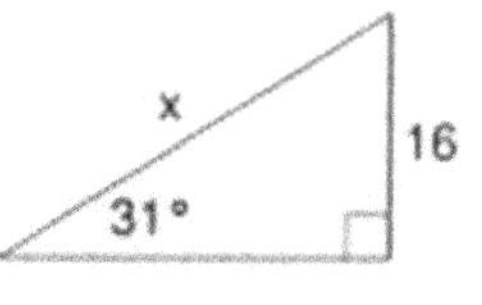

Arc Length and Sector Area

Step-by-step guide:

✓ To find a sector of a circle, use this formula: $Area\ of\ a\ sector = \pi r^2(\frac{\theta}{360})$, r is the radius of the circle and θ is the central angle of the sector.

✓ To find the arc of a sector of a circle, use this formula: $Arc\ of\ a\ sector = (\frac{\theta}{180})\pi r$

Examples:

1) ***Find the length of the arc. Round your answers to the nearest tenth.*** $(\pi = 3.14)$
$r = 28$ cm, $\theta = 45°$

Use this formula: $length\ of\ a\ sector = (\frac{\theta}{180})\pi r$

$length\ of\ a\ sector = \left(\frac{45}{180}\right)\pi(28) = \left(\frac{1}{4}\right)\pi(28) = 7 \times 3.14 \cong 22\ cm$

2) ***Find the area of the sector.*** $r = 5$ ft, $\theta = 80°$

Use this formula: $Area\ of\ a\ sector = \pi r^2(\frac{\theta}{360})$

$Area\ of\ a\ sector = \pi r^2\left(\frac{\theta}{360}\right) = (3.14)(5^2)\left(\frac{80}{360}\right) \cong 17.4$

✍ ***Find the length of each arc. Round your answers to the nearest tenth.***

1) $r = 22$ ft, $\theta = 60°$
2) $r = 12\ m$, $\theta = 85°$

✍ ***Find area of each sector. Round your answers to the nearest tenth.*** $(\pi = 3.14)$

3)

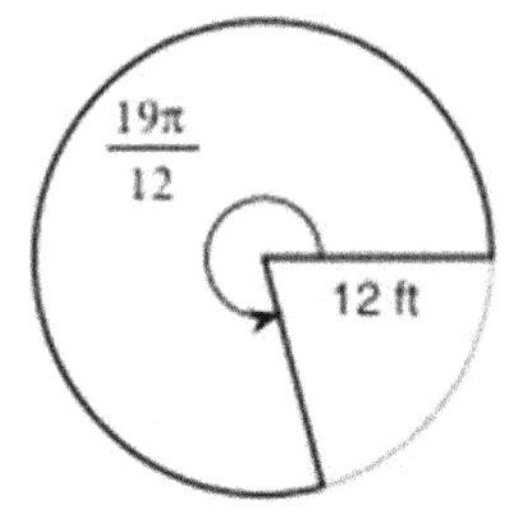

4)

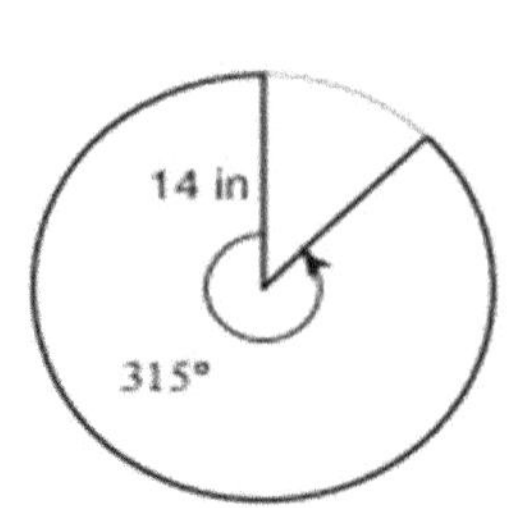

5)

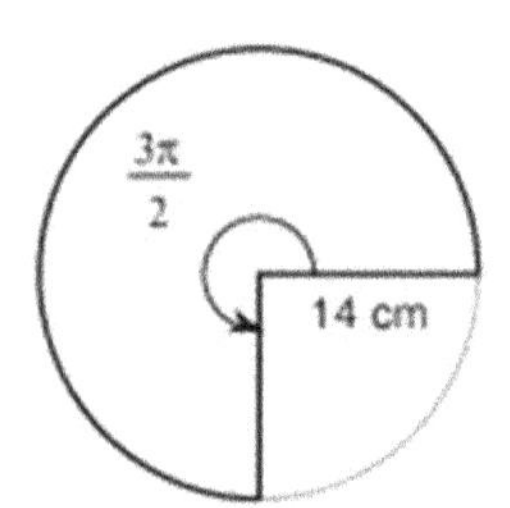

6)

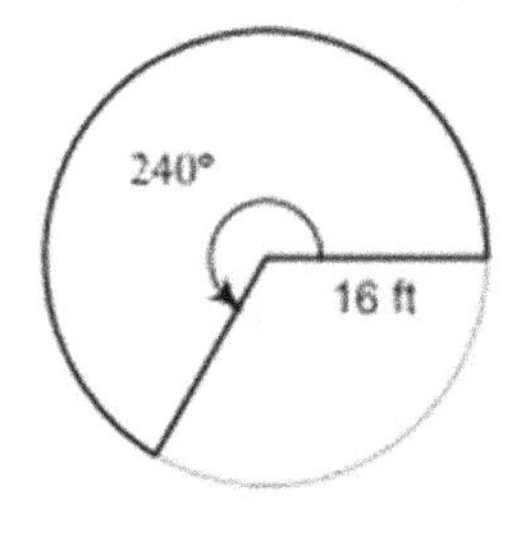

Answers of Worksheets – Chapter 13

Trig ratios of general angles

1) $-\frac{\sqrt{3}}{2}$
2) $\frac{1}{2}$
3) $\frac{\sqrt{2}}{2}$
4) -1
5) $\frac{\sqrt{3}}{2}$
6) $\frac{1}{2}$
7) Undefined
8) 0
9) Undefined
10) $-\sqrt{3}$
11) -2
12) $undefined$

Co–Terminal Angles and Reference Angles

1) $280°$
2) $280°$
3) $285°$
4) $30°$
5) $\frac{7\pi}{4}$
6) $\frac{5\pi}{12}$
7) $\frac{\pi}{18}$
8) $\frac{5\pi}{3}$

Angles and Angle Measure

1) $-\frac{7\pi}{9}$
2) $\frac{16\pi}{9}$
3) $\frac{7\pi}{6}$
4) $\frac{13\pi}{3}$
5) $-\frac{19\pi}{18}$
6) $\frac{23\pi}{12}$
7) $6°$
8) $144°$
9) $70°$
10) $36°$
11) $-225°$
12) $840°$

Evaluating Trigonometric Expression

1) $-\frac{\sqrt{3}}{3}$
2) $-\sqrt{3}$
3) $\frac{\sqrt{2}}{2}$
4) $-\frac{1}{2}$
5) $-\frac{1}{2}$
6) $\sqrt{3}$

Missing sides and angles of a right triangle

1) $\frac{4}{3}$
2) $\frac{3}{5}$
3) 31.4
4) 7.0
5) 16.2
6) 31.1

Arc length and sector area

1) $23\ ft$

2) $17.8\ m$

3) $358\ ft^2$

4) $538.5\ in^2$

5) $461.6\ cm^2$

6) $535.9\ ft^2$

Chapter 14: Sequences and Series

Topics that you'll learn in this chapter:

- ✓ Arithmetic Sequences
- ✓ Geometric Sequences
- ✓ Finite Geometric Series
- ✓ Infinite Geometric Series

Arithmetic Sequences

Step-by-step guide:

✓ A sequence of numbers such that the difference between the consecutive terms is constant is called arithmetic sequence. For example, the sequence 6, 8, 10, 12, 14, ... is an arithmetic sequence with common difference of 2.

✓ To find any term in an arithmetic sequence use this formula: $x_n = a + d(n - 1)$
a = the first term, d = the common difference between terms, n = number of items

Examples:

1) Given the first term and the common difference of an arithmetic sequence find the first five terms. $a_1 = 24, d = 2$

Use arithmetic sequence formula: $x_n = a + d(n - 1)$,

If $n = 1$ then: $x_1 = 22 + 2(1) \rightarrow x_1 = 24$

First Five Terms: $24, 26, 28, 30, 32$

2) Find the first five terms of the sequence. $a_{17} = 152, d = 4$

First, we need to find a_1 or a. Use arithmetic sequence formula: $x_n = a + d(n - 1)$

If $a_{17} = 152$, then $n = 17$. Rewrite the formula and put the values provided:

$x_n = a + d(n - 1) \rightarrow 152 = a + 4(17 - 1) = a + 64$, now solve for a.

$152 = a + 64 \rightarrow a = 152 - 64 = 88$,

First Five Terms: $88, 92, 96, 100, 104$

Given a term in an arithmetic sequence and the common difference find the first five terms of the sequence.

1) $a_{36} = -276, d = -7$
2) $a_{37} = 249, d = 8$
3) $a_{38} = -53.2, d = -1.1$
4) $a_{40} = -1{,}191, d = -30$

Given a term in an arithmetic sequence and the common difference find the three terms in the sequence after the last one given.

5) $a_{22} = -44, d = -2$
6) $a_{12} = 28.6, d = 1.8$
7) $a_{18} = 27.4, d = 1.1$
8) $a_{21} = -1.4, d = 0.6$

Geometric Sequences

Step-by-step guide:

- ✓ It is a sequence of numbers where each term after the first is found by multiplying the previous item by the common ratio, a fixed, non-zero number. For example, the sequence 2, 4, 8, 16, 32, … is a geometric sequence with common ratio of 2.
- ✓ To find any term in a geometric sequence use this formula: $x_n = ar^{(n-1)}$

  a = the first term, r = the common ratio, n = number of items

Examples:

1) ***Given the first term and the common ratio of a geometric sequence find the first five terms of the sequence.*** $a_1 = 0.8, r = -5$

   Use geometric sequence formula: $x_n = ar^{(n-1)} \rightarrow x_n = 0.8\,.(-5)^{n-1}$

   If $n = 1$ then: $x_1 = 0.8\,.(-5)^{1-1} = 0.8\,(1) = 0.8$, First Five Terms: $0.8, -4, 20, -100, 500$

2) ***Given two terms in a geometric sequence find the 8th term.*** $a_3 = 12$ ***and*** $a_5 = 48$

   Use geometric sequence formula: $x_n = ar^{(n-1)} \rightarrow a_3 = ar^{(3-1)} = ar^2 = 12$

   $$x_n = ar^{(n-1)} \rightarrow a_5 = ar^{(5-1)} = ar^4 = 48$$

   Now divide a_5 by a_3. Then: $\frac{a_5}{a_3} = \frac{ar^4}{ar^2} = \frac{48}{12}$, Now simplify: $\frac{ar^4}{ar^2} = \frac{48}{12} \rightarrow r^2 = 4 \rightarrow r = 2$

   We can find a now: $ar^2 = 12 \rightarrow a(2^2) = 12 \rightarrow a = 3$

   Use the formula to find the 8th term: $x_n = ar^{(n-1)} \rightarrow a_8 = (3)(2)^8 = 3(256) = 768$

✍ ***Given the recursive formula for a geometric sequence find the common ratio and the first five terms.***

1) $a_n = a_{n-1}\,.2, a_1 = 2$
2) $a_n = a_{n-1}\,.-3, a_1 = -3$
3) $a_n = a_{n-1}\,.5, a_1 = 2$
4) $a_n = a_{n-1}\,.3, a_1 = -3$

✍ ***Given two terms in a geometric sequence find the 8th term.***

5) $a_4 = 12$ and $a_5 = -6$
6) $a_5 = 768$ and $a_2 = 12$

Finite Geometric Series

Step-by-step guide:

- ✓ The sum of a geometric series is finite when the absolute value of the ratio is less than 1.
- ✓ Finite Geometric Series formula: $S_n = \sum_{i=1}^{n} ar^{i-1} = a_1(\frac{1-r^n}{1-r})$

Examples:

Evaluate each geometric series described.

1) $\sum_{n=1}^{7} 2^{n-1}$

Use this formula: $S_n = \sum_{i=1}^{n} ar^{i-1} = a_1\left(\frac{1-r^n}{1-r}\right) \rightarrow \sum_{n=1}^{7} 2^{n-1} = 1\left(\frac{1-2^7}{1-2}\right)$

$1\left(\frac{1-2^7}{1-2}\right) = 1\left(\frac{1-128}{1-2}\right) = \left(\frac{-127}{-1}\right) = 127$

2) $\sum_{n=1}^{4} -5^{n-1}$

Use this formula: $S_n = \sum_{i=1}^{n} ar^{i-1} = a_1\left(\frac{1-r^n}{1-r}\right) \rightarrow \sum_{n=1}^{4} -5^{n-1} = (-1)\left(\frac{1-5^4}{1-5}\right)$

$(-1)\left(\frac{1-5^4}{1-5}\right)(-1)\left(\frac{1-625}{1-5}\right) = (-1)\left(\frac{-624}{-4}\right) = (-1)\left(\frac{624}{4}\right) = -156$

Evaluate each geometric series described.

1) $a_1 = -1, r = 4, n = 8$________

2) $a_1 = -2, r = -3, n = 9$________

3) $\sum_{n=1}^{8} 2 \,.\, (-2)^{n-1}$________

4) $\sum_{n=1}^{9} 4 \,.\, 3^{n-1}$ ________

5) $\sum_{n=1}^{10} 4 \,.\, (-3)^{n-1}$________

6) $\sum_{m=1}^{9} -2^{m-1}$________

7) $\sum_{m=1}^{8} 3 \,.\, 5^{m-1}$________

8) $\sum_{k=1}^{7} 2 \,.\, 5^{k-1}$________

Infinite Geometric Series

Step-by-step guide:

- ✓ Infinite Geometric Series: The sum of a geometric series is infinite when the of the ratio is more than 1.
- ✓ Infinite Geometric Series formula: $S = \sum_{i=0}^{\infty} a_i r^i = \frac{a_1}{1-r}$

Examples:

1) ***Evaluate infinite geometric series described.*** $\sum_{i=1}^{\infty} 8^{i-1}$

Use this formula: $\sum_{i=0}^{\infty} a_i r^i = \frac{a_1}{1-r} \rightarrow \sum_{i=1}^{\infty} 8^{i-1} = \frac{1}{1-8} = \frac{1}{-7} = -\frac{1}{7}$

2) ***Evaluate infinite geometric series described.*** $\sum_{k=1}^{\infty} (\frac{1}{2})^{k-1}$

Use this formula: $\sum_{i=0}^{\infty} a_i r^i = \frac{a_1}{1-r} \rightarrow \sum_{k=1}^{\infty} (\frac{1}{2})^{k-1} = \frac{1}{1-\frac{1}{2}} = \frac{1}{\frac{1}{2}} = 2$

Evaluate each infinite geometric series described.

1) $a_1 = 3, \mathrm{r} = -\frac{1}{5}$

2) $a_1 = 1, \mathrm{r} = -3$

3) $a_1 = 1, \mathrm{r} = -4$

4) $a_1 = 3, \mathrm{r} = \frac{1}{2}$

5) $1 + 0.5 + 0.25 + 0.125 + \cdots$

6) $81 - 27 + 9 - 3$ ...,

7) $1 - 0.6 + 0.36 - 0.216$ ...,

8) $3 + \frac{9}{4} + \frac{27}{16} + \frac{81}{64}$ ...,

9) $\sum_{k=1}^{\infty} 4^{k-1}$

10) $\sum_{i=1}^{\infty} (\frac{1}{3})^{i-1}$

11) $\sum_{k=1}^{\infty} (-\frac{1}{3})^{k-1}$

12) $\sum_{n=1}^{\infty} 16(\frac{1}{4})^{n-1}$

Answers of Worksheets – Chapter 14

Arithmetic Sequences

1) First five terms: $-31, -38, -45, -52, -59$
2) First five terms: $-39, -31, -23, -15, -7$
3) First five terms: $-12.5, -13.6, -14.7, -15.8, -16.9$
4) First five terms: $-21, -51, -81, -111, -141$
5) Next 3 terms: $-46, -48, -50$
6) Next 3 terms: $30.4, 32.2, 34$
7) Next 3 terms: $28.5, 29.6, 30.7$
8) Next 3 terms: $-0.8, -0.2, 0.4$

Geometric Sequences

1) Common Ratio: $r = 2$, First five terms: $2, 4, 8, 16, 32$
2) Common Ratio: $r = -3$, First five terms: $-3, 9, -27, 81, -243$
3) Common Ratio: $r = 5$, First five terms: $2, 10, 50, 250, 1{,}250$
4) Common Ratio: $r = 3$, First five terms: $-3, -9, -27, -81, -243$
5) $a_8 = \frac{3}{4}, a_1 = -96$
6) $a_8 = 49{,}152.4, a_1 = 3$

Finite Geometric

1) $-21{,}845$
2) $-9{,}842$
3) -170
4) $39{,}364$
5) -59048
6) 171
7) $292{,}968$
8) $39{,}062$

Infinite Geometric

1) $\frac{5}{2}$
2) Infinite
3) Infinite
4) 6
5) 2
6) $\frac{243}{4}$
7) 0.625
8) 12
9) Infinite
10) $\frac{3}{2}$
11) $\frac{3}{4}$
12) $\frac{64}{3}$

"Effortless Math" Publications

Effortless Math authors' team strives to prepare and publish the best quality TSI Mathematics learning resources to make learning Math easier for all. We hope that our publications help you learn Math in an effective way and prepare for the TSI test.

We all in Effortless Math wish you good luck and successful studies!

Effortless Math Authors

www.EffortlessMath.com

... So Much More Online!

- ✓ FREE Math lessons
- ✓ More Math learning books!
- ✓ Mathematics Worksheets
- ✓ Online Math Tutors

Need a PDF version of this book?

Please visit www.EffortlessMath.com

CPSIA information can be obtained
at www.ICGtesting.com
Printed in the USA
BVHW012105110520
579543BV00007B/78